NW

FRONTIERS OF SCIENCE

EARTH SCIENCES

FRONTIERS OF SCIENCE

EARTH SCIENCES

Notable Research and Discoveries

KYLE KIRKLAND, PH.D.

Facts On File
An imprint of Infobase Publishing

EARTH SCIENCES: Notable Research and Discoveries

Facts On File, Inc.
An imprint of Infobase Publishing
132 West 31st Street
New York NY 10001

Library of Congress Cataloging-in-Publication Data

Kirkland, Kyle.
　Earth sciences: notable research and discoveries / Kyle Kirkland.
　　p. cm.—(Frontiers of science)
　Includes bibliographical references and index.
　ISBN 978-0-8160-7442-6
　1. Geology. 2. Earth sciences. I. Title

　QE501.K527 2010
　550—dc22　　　2009020269

Facts On File books are available at special discounts when purchased in bulk quantities for businesses, associations, institutions, or sales promotions. Please call our Special Sales Department in New York at (212) 967-8800 or (800) 322-8755.

You can find Facts On File on the World Wide Web at http://www.factsonfile.com

Excerpts included herewith have been reprinted by permission of the copyright holders; the author has made every effort to contact copyright holders. The publishers will be glad to rectify, in future editions, any errors or omissions brought to their notice.

Text design by Kerry Casey
Illustrations by Sholto Ainslie and Dale Williams
Photo research by Tobi Zausner, Ph.D.
Composition by Mary Susan Ryan-Flynn
Cover printed by Bang Printing, Inc., Brainerd, Minn.
Book printed and bound by Bang Printing, Inc., Brainerd, Minn.
Date printed: March 2010
Printed in the United States of America

10 9 8 7 6 5 4 3 2 1

This book is printed on acid-free paper.

CONTENTS

PREFACE

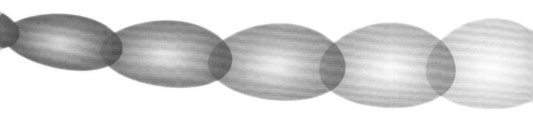

Discovering what lies behind a hill or beyond a neighborhood can be as simple as taking a short walk. But curiosity and the urge to make new discoveries usually require people to undertake journeys much more adventuresome than a short walk, and scientists often study realms far removed from everyday observation—sometimes even beyond the present means of travel or vision. Polish astronomer Nicolaus Copernicus's (1473–1543) heliocentric (Sun-centered) model of the solar system, published in 1543, ushered in the modern age of astronomy more than 400 years before the first rocket escaped Earth's gravity. Scientists today probe the tiny domain of atoms, pilot submersibles into marine trenches far beneath the waves, and analyze processes occurring deep within stars.

Many of the newest areas of scientific research involve objects or places that are not easily accessible, if at all. These objects may be trillions of miles away, such as the newly discovered planetary systems, or they may be as close as inside a person's head; the brain, a delicate organ encased and protected by the skull, has frustrated many of the best efforts of biologists until recently. The subject of interest may not be at a vast distance or concealed by a protective covering, but instead it may be removed in terms of time. For example, people need to learn about the evolution of Earth's weather and climate in order to understand the changes taking place today, yet no one can revisit the past.

Frontiers of Science is an eight-volume set that explores topics at the forefront of research in the following sciences:

- biological sciences
- chemistry

- computer science
- Earth science
- marine science
- physics
- space and astronomy
- weather and climate

The set focuses on the methods and imagination of people who are pushing the boundaries of science by investigating subjects that are not readily observable or are otherwise cloaked in mystery. Each volume includes six topics, one per chapter, and each chapter has the same format and structure. The chapter provides a chronology of the topic and establishes its scientific and social relevance, discusses the critical questions and the research techniques designed to answer these questions, describes what scientists have learned and may learn in the future, highlights the technological applications of this knowledge, and makes recommendations for further reading. The topics cover a broad spectrum of the science, from issues that are making headlines to ones that are not as yet well known. Each chapter can be read independently; some overlap among chapters of the same volume is unavoidable, so a small amount of repetition is necessary for each chapter to stand alone. But the repetition is minimal, and cross-references are used as appropriate.

Scientific inquiry demands a number of skills. The National Committee on Science Education Standards and Assessment and the National Research Council, in addition to other organizations such as the National Science Teachers Association, have stressed the training and development of these skills. Science students must learn how to raise important questions, design the tools or experiments necessary to answer these questions, apply models in explaining the results and revise the model as needed, be alert to alternative explanations, and construct and analyze arguments for and against competing models.

Progress in science often involves deciding which competing theory, model, or viewpoint provides the best explanation. For example, a major issue in biology for many decades was determining if the brain functions as a whole (the holistic model) or if parts of the brain carry out specialized functions (functional localization). Recent developments in brain imaging resolved part of this issue in favor of functional localization by showing that specific regions of the brain are more active during certain tasks. At the same time, however, these experiments have raised other questions that future research must answer.

The logic and precision of science are elegant, but applying scientific skills can be daunting at first. The goals of the Frontiers of Science set are to explain how scientists tackle difficult research issues and to describe recent advances made in these fields. Understanding the science behind the advances is critical because sometimes new knowledge and theories seem unbelievable until the underlying methods become clear. Consider the following examples. Some scientists have claimed that the last few years are the warmest in the past 500 or even 1,000 years, but reliable temperature records date only from about 1850. Geologists talk of volcano hot spots and plumes of abnormally hot rock rising through deep channels, although no one has drilled more than a few miles below the surface. Teams of neuroscientists—scientists who study the brain—display images of the activity of the brain as a person dreams, yet the subject's skull has not been breached. Scientists often debate the validity of new experiments and theories, and a proper evaluation requires an understanding of the reasoning and technology that support or refute the arguments.

Curiosity about how scientists came to know what they do—and why they are convinced that their beliefs are true—has always motivated me to study not just the facts and theories but also the reasons why these are true (or at least believed). I could never accept unsupported statements or confine my attention to one scientific discipline. When I was young, I learned many things from my father, a physicist who specialized in engineering mechanics, and my mother, a mathematician and computer systems analyst. And from an archaeologist who lived down the street, I learned one of the reasons why people believe Earth has evolved and changed—he took me to a field where we found marine fossils such as shark's teeth, which backed his claim that this area had once been under water! After studying electronics while I was in the air force, I attended college, switching my major a number of times until becoming captivated with a subject that was itself a melding of two disciplines—biological psychology. I went on to earn a doctorate in neuroscience, studying under physicists, computer scientists, chemists, anatomists, geneticists, physiologists, and mathematicians. My broad interests and background have served me well as a science writer, giving me the confidence, or perhaps I should say chutzpah, to write a set of books on such a vast array of topics.

Seekers of knowledge satisfy their curiosity about how the world and its organisms work, but the applications of science are not limited to intellectual achievement. The topics in Frontiers of Science affect society on a multitude of levels. Civilization has always faced an uphill battle to procure scarce resources, solve technical problems, and maintain order. In modern times, one of the most important resources is energy, and the physics of fusion potentially offers a nearly boundless supply. Technology makes life easier and solves many of today's problems, and nanotechnology may extend the range of devices into extremely small sizes. Protecting one's personal information in transactions conducted via the Internet is a crucial application of computer science.

But the scope of science today is so vast that no set of eight volumes can hope to cover all of the frontiers. The chapters in Frontiers of Science span a broad range of each science but could not possibly be exhaustive. Selectivity was painful (and editorially enforced) but necessary, and in my opinion, the choices are diverse and reflect current trends. The same is true for the subjects within each chapter—a lot of fascinating research did not get mentioned, not because it is unimportant, but because there was no room to do it justice.

Extending the limits of knowledge relies on basic science skills as well as ingenuity in asking and answering the right questions. The 48 topics discussed in these books are not straightforward laboratory exercises but complex, gritty research problems at the frontiers of science. Exploring uncharted territory presents exceptional challenges but also offers equally impressive rewards, whether the motivation is to solve a practical problem or to gain a better understanding of human nature. If this set encourages some of its readers to plunge into a scientific frontier and conquer a few of its unknowns, the books will be worth all the effort required to produce them.

ACKNOWLEDGMENTS

Thanks go to Frank K. Darmstadt, executive editor at Facts On File, and his staff for all their hard work, which I admit I sometimes made a little bit harder. Thanks also to Tobi Zausner for researching and locating so many great photographs. I also appreciate the time and effort of a large number of researchers who were kind enough to pass along a research paper or help me track down some information.

INTRODUCTION

The flat Earth hypothesis tends to draw a giggle from students these days. Everyone knows the world is round—photographs of Earth taken from space show a bluish sphere partially obscured with white clouds. Yet in the past beliefs in a flat Earth were not as ridiculous as they appear today.

Consider the view from the surface. In most places, Earth looks flat. To a person who has never seen the view from space, it is reasonable to suppose that what seems to be true locally—flatness—is true everywhere. Many intelligent people have held this view. The grandfather of the author of this book was one of them; he was a successful farmer who knew more about soil and how to work the land than most geology professors, but he had difficulty accepting the notion that the world was anything other than what it appeared to his eyes. No argument could ever quite convince this skeptic.

Earth Sciences: Notable Research and Discoveries, one volume in the multivolume Frontiers of Science set, is about explorers and scientists who venture into the unknown frontiers of Earth science—and quite often run into things they do not expect. Earth is a large, complicated object to study and home to environments as different as rain forests and deserts. The interactions of the planet's various components, such as the atmosphere, oceans, land, and the rocks and metals of its interior, produce a bewildering array of phenomena. Many of these phenomena have a strong impact on people's lives, although the realm of human society does not generally extend beyond Earth's surface.

People who probe the frontiers of science get a more complete view of their world. In the third century B.C.E., the Greek philosopher and mathematician Eratosthenes (ca. 276–194 B.C.E.) calculated Earth's diameter by

Earth, as seen from space *(Johnson Space Center/NASA)*

observing the position of the Sun at different places on Earth's surface. Eratosthenes knew that at noon on the summer solstice—the longest day of the year—the Sun appeared directly overhead in Syene (now Aswan), an Egyptian city. In Alexandria, a city north of Syene, Eratosthenes measured the position of the Sun on the summer solstice and found it made an angle of slightly more than seven degrees from the point directly overhead. This angle is about 1/50 of a full circle (which is 360 degrees). Eratosthenes reasoned that if Earth was a sphere, then his measurement indicated that the distance from Alexandria to Syene was about 1/50 the circumference of the planet. By estimating this distance and multiplying by 50, Eratosthenes calculated the value of Earth's circumference to be (in modern units) about 24,500 miles (39,500 km), a figure quite close to the modern value, 24,900 miles (40,161 km).

The Italian navigator Christopher Columbus (1451–1506) believed the world was round and sailed west from Spain in an attempt to reach the East Indies, the site of valuable trade that at the time could only be reached by arduous land journeys across eastern Europe and Asia. (In the 15th century Ottoman Turks had cut off most of the land routes.) Skeptics scoffed at Columbus's idea, but not because they thought the world was flat and Columbus would sail off the edge. Instead, many people thought the journey was too long to be successfully navigated, as suggested by the calculations of Eratosthenes. They were right—but nobody knew about America until Columbus ran into it and mistook it for his destination.

Scientists have made much progress in recent years, but there are still many frontiers awaiting the calculations of an Eratosthenes, the visit of a Columbus, or the invention of a device such as a satellite to get a better view. Each chapter of this book explores one of these frontiers. Reports published in journals, presented at conferences, and described in press releases illustrate the kind of research problems of interest in Earth science and how scientists attempt to solve them. This book summarizes a selection of these reports—there is room for only a fraction of them—that offer students and other readers insight into the methods and applications of Earth science.

Students need to keep up with the latest developments in these quickly moving fields of research, but they have difficulty finding a source that explains the basic concepts while discussing the background and context essential for the big picture. This book describes the evolution of each of the six main topics it covers and explains the problems that researchers are currently investigating, as well as the methods they are developing to solve them.

Earth's interior is a fascinating frontier—and far removed from the sight and reach of scientists eager to learn what lies hidden below. Chapter 1 recounts the efforts to plumb Earth's depths. Digging and drilling form only part of the solution, since no one can drill very far below the surface and certainly nowhere close to Earth's center. Instead of paying a visit in person, scientists study the depths of the planets by using technology to extend their vision. The recording and analysis of waves rumbling through Earth's innards allow researchers to explore these otherwise inaccessible places.

Sometimes complexity lies masked behind the simplest phenomenon. To sailors such as Columbus, compasses were simple yet essential instruments, indicating direction as the explorers navigated the vast oceans. The mechanism that makes compasses work was a vague subject until 1600, when the British physicist and physician William Gilbert (1544–1603) studied the magnetic properties of Earth. But later, after more detailed studies, researchers discovered that Earth's magnetism is not quite as simple as it seems to be. As discussed in chapter 2, the pursuit of this phenomenon has taken scientists to the very core of the planet.

One of the reasons why people cannot dig into the ground very deeply is that sooner or later they will encounter temperatures high enough to melt their equipment. The molten interior makes its presence felt in other ways, sometimes reaching the surface and erupting from volcanoes. Scientists in the 20th century discovered that most volcanoes are located around cracks between the large, moving plates that cover Earth's surface. But other volcanoes, such as those in the Hawaiian Islands, are far removed from any plate boundary and therefore require an alternative explanation. Chapter 3 explains how geologists are rethinking their ideas of Earth's structure and dynamics in the quest for the correct explanation.

The heat lying just below the surface has given researchers and engineers other ideas as well. Heat is energy, and society today needs a lot of energy to power technological devices and transport cargoes and people. Fuels such as oil and coal deliver much of this energy, but their limited supply and the pollution accompanying their use motivate a search for alternative resources. One alternative is the heat of Earth's interior. In chapter 4, the methods people have employed to extract this energy are discussed, along with research aimed at improving these methods and multiplying the yield.

Another critical resource is water. Earth has a lot of it, but most of it is salt water in the ocean. Freshwater suitable for drinking is scarce in many parts of the world, and, as populations grow, shortages are becoming an increasing problem even for the more technological advanced countries. To improve the situation, researchers must achieve a better understanding of how water is distributed in the planet, how it flows, and how to make the best use of what is available. Chapter 5 discusses this important frontier of Earth science.

Chapter 6 finishes the book with a topic that is a perennial favorite of students. Earthquakes are greatly feared and aptly so; they also instill a kind of awe. A person standing on the ground usually gets a comforting impression of solidity, but this impression is obliterated during the shaking and rolling of a major earthquake. Damage and casualties can be severe. Scientists have learned to forecast the probability of a major earthquake during a certain interval of time—usually decades—but people would like to do better in order to provide a more precise warning and reduce casualties. Specific predictions of when and where the next earthquake will strike have so far eluded the ability of the best researchers, but intrepid scientists at the frontiers of Earth science are continuing to try.

These chapters do not exhaust the list of frontiers of Earth science, but they offer a broad spectrum of topics, each of which has the potential for significant breakthroughs. Earth is home to 6 billion people and a plethora of other forms of life. It is, at least presently, the only home humans have, so it is vital to the health and welfare of society. This book will help readers understand how Earth scientists conduct research and what this research might bring in the future.

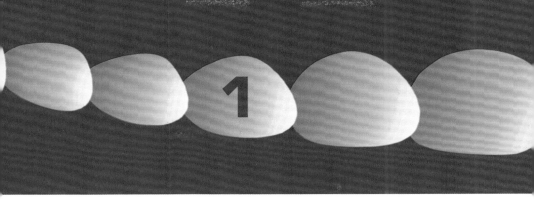

EXPLORING
EARTH'S DEPTHS

Geologists of the 19th century faced an interesting puzzle—Earth seemed to have a great deal more mass than it should have had. At the close of the 18th century, the British scientist Henry Cavendish (1731–1810) calculated the planet's density—its mass divided by the volume. Cavendish accomplished this feat by precisely measuring the force of gravity between two objects and then applying Newton's law of universal gravitation. The value Cavendish reported in 1798 for Earth's density was about 5.5 times that of water, which is quite close to the value accepted by modern scientists. Yet this posed a difficult question for geologists. Rocks on Earth's surface are only about 2.7 times denser than water, and no one knew how Earth's density could be twice as large as the rocks that compose it!

One possible solution involves the concept of pressure. Ocean divers experience increased pressure as they descend beneath the surface because of the weight of the water pressing down on them. In the same way, rocks on the surface press down on rocks beneath them, increasing the pressure and causing subsurface objects to squeeze into smaller space, which will increase the density—the same mass in a smaller volume results in a higher density.

The German scientist Emil Wiechert (1861–1928) offered a bolder hypothesis in 1896. He rejected the notion of pressure as a cause for the higher density. As Stephen G. Brush wrote in his 1996 book *Nebulous Earth,* "Supposing (in accordance with 19th-century ideas) that the molecules in a solid are already very close together at low pressures, Wiechert argued that density cannot be increased very much by compression; hence, the density difference must be attributed to a difference in chemical composition rather than merely to pressure." Instead of a completely rocky interior, Wiechert proposed that Earth's core is made of metal. Mixtures of iron and nickel that are commonly found in meteorites have a density of up to about eight times that of water. Although no miner can drill far enough into the planet to reach this metallic core, its existence would explain why Earth has a much higher overall density than the surface rocks.

With no means of inspecting a hypothetical metallic core, confirming this hypothesis would appear impossible. This kind of problem is prevalent in geology. Much of the planet lies hidden and inaccessible beneath its surface, and people have thus far observed only about 1 percent of Earth. But science has expanded humanity's reach and vision in a variety of ways. In the case of exploring Earth's depths, geologists have devised techniques to see where people cannot yet go. Some of these techniques include the study of waves so powerful that they shake the ground—these are the *seismic waves* of earthquakes. This chapter discusses these techniques and how researchers have used them to support Wiechert's ideas and to make new discoveries about what lies beneath Earth's surface.

INTRODUCTION

The radius of Earth is about 3,950 miles (6,370 km). This means that the distance between Earth's surface and its center is approximately the same as the distance between Washington, D.C., and Paris, France. Although it might not sound very far, digging or drilling through the ground for more than a small fraction of this distance is not possible.

Knowledge of Earth's interior is important for many reasons, however. Many essential resources are buried belowground, including metals needed for industry and oil, coal, and natural gas that are extracted for energy. These resources must be found and mined, bringing them up to the surface. Some mines are open pits—large holes or cavities created by removing the topsoil—and some mines consist of small vertical shafts constantly under threat of cave-ins. The deepest mines in the world are the gold mines of South Africa, some of which reach a

Gold mine in Alaska *(Natalia Bratslavksy/iStockphoto)*

depth of 12,800 feet (3,900 m), or 2.4 miles (3.9 km). Finding hidden resources is not easy. Geologists determine the most promising sites so that miners do not waste time and effort hunting in the wrong spot. For example, oil deposits are most often located around *sedimentary rock,* a type of rock formed when sediments or layers of sand, mud, or the calcium carbonate shells of organisms become buried and compacted.

Understanding Earth's interior is also important because it helps scientists understand the planet, as well as the changes in climate and environment the planet is currently experiencing. A fuller knowledge of Earth will help scientists understand how the solar system formed and the nature of other solar systems and planets throughout the galaxy (as of May 2009, astronomers have detected 346 planets outside the solar system). In addition, Earth's depths constitute an exciting frontier of science, beckoning explorers who have a passionate desire to learn what lies hidden beneath their feet.

The uppermost layer of Earth, called the *crust,* contains the mountains, plains, and deserts of the continents and the seafloor. Most of the rocks of the crust are composed of silicates—compounds containing the elements silicon (Si) and oxygen (O), such as silica (SiO_2), a molecule which consists of one silicon atom and two oxygen atoms. Sand and quartz are common examples. Another common silicate known as olivine contains iron and magnesium along with silicon and oxygen. In terms of chemical elements, the weight of Earth's crust is about 46 percent oxygen, 28 percent silicon, 8 percent aluminum, 6 percent iron, 4 percent magnesium, and a small percentage of other elements.

Major features such as mountains do not seem to change much in a human lifetime, yet Earth is a dynamic place. The top of Mount Everest, which soars more than 29,030 feet (8,850 m) above sea level, is rich in

Despite seemingly permanent features, such as Mount Rushmore in South Dakota, Earth is constantly, albeit slowly, changing. *(William Walsh/ iStockphoto)*

limestone—a sedimentary rock—that contains marine fossils and was once under water! In 1912 the German researcher Alfred Wegener (1880–1930) noticed that the coasts of continents such as Africa and South America seemed to fit together and displayed remarkable similarities in the kind of fossils they contained, as if these now-separated continents were once adjoined. He proposed the notion of continental drift and hypothesized that continents had once been joined. Wegener had a difficult time convincing people that something as massive as a continent moves, and he was wrong, as it turned out, in some of his ideas—Wegener was unable to propose a viable mechanism by which continents move, and he incorrectly believed continents float across oceans. But anyone who has ever lived through an earthquake knows the ground can certainly move.

SEISMIC WAVES

Wiechert, Wegener, and other researchers encouraged their colleagues to reexamine assumptions about the dynamics and structure of Earth's interior. But ideas alone are not sufficiently convincing. Scientific evidence that supports a hypothesis or a particular point of view is essential before the scientific community is willing to accept an idea. Although obtaining evidence on the nature of Earth's depths or on any other location where it is not yet possible to venture is extremely difficult, geologists of the early 20th century began using seismic waves as their eyes into the planet's interior. These waves continue to be the most important tool for these studies today.

Waves are important in many branches of science, especially the study of sound and light, both of which behave (at least under certain conditions) as waves. A wave is a vibration or disturbance that propagates across space or in a material such as water or air. To make a wave, something has to fluctuate—electromagnetic fields in the case of light, air pressure in the case of sound, or water in the case of sea or lake waves—and it is this fluctuation that propagates. For instance, a stone dropped in a pond will create ripples spreading out from the point at which the stone fell. The fall of the stone created a disturbance that moved the water in the small region surrounding the impact zone, and these water molecules pushed against their neighbors, and so on, propagating the disturbance throughout the pond.

Disturbances can propagate in several different ways. A transverse wave propagates in a direction perpendicular (at a 90 degree angle) to the vibrations or oscillations, as illustrated in the bottom of the figure on page 7. Light waves are examples of transverse waves. Inside solid materials, the side-to-side oscillation (with respect to the direction of travel) is associated with a kind of force known as shear stress, so these waves are sometimes called *shear waves,* a term geologists often use because many of the waves they study travel through solids. The top of the figure illustrates another kind of wave, called a longitudinal wave, which propagates in the same direction as the vibrations. Sound waves are longitudinal waves, since a sound wave consists of a compression propagating through air, water, or some other material, caused by molecules moving toward (and then away) from each other in the same direction that the wave propagates. The compression gives these waves an alternative name—*compression waves.*

Wave behavior is critical in optics (the study and use of light) and acoustics (the study and use of sound). Camera lenses form images on film or digital sensors by bending and focusing light, and eyeglasses and contact lenses perform a similar service for people whose vision would otherwise be blurry. The focusing is due to refraction—the bending of the wave when passing from one substance to another. For instance, when a light wave passes from air into the transparent glass of a lens, light changes speed, which causes its path to bend, or refract. Another property of waves that occurs at a boundary between two different substances is reflection—some of the motion is sent back. For example, the glass of a window transmits a lot of light but also reflects some of it, so an observer looking through a window can see outside but may also notice his or her reflection in the glass.

The speed of waves is also crucial. Waves travel at a specific speed in the material, or medium, through which the disturbance is propagating. In general, compression waves travel faster in a medium that resists compression. For example, sound waves travel faster in the denser air at

(opposite page) Compression waves consist of contractions and expansions in the same direction (longitudinally) as the propagation of the wave. Shear or transverse waves consist of up-and-down motions perpendicular to the wave's propagation.

Earth's surface than the thinner air high in the atmosphere. Chuck Yeager, who in 1947 made the first documented flight exceeding the speed of sound, flew at an altitude of about 45,000 feet (13.7 km), where the

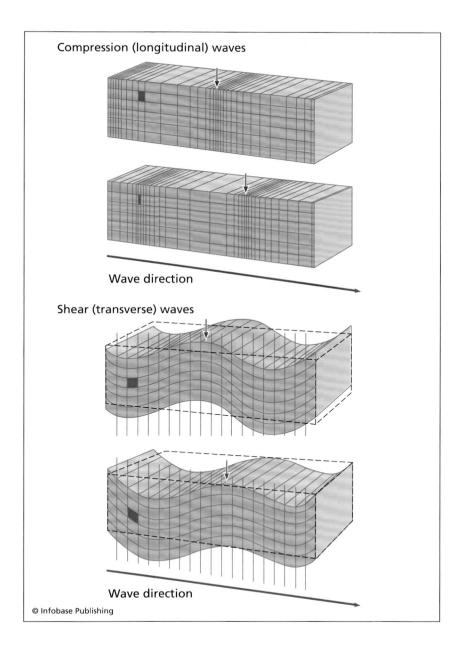

Compression (longitudinal) waves

Wave direction

Shear (transverse) waves

Wave direction

© Infobase Publishing

Seismic recording equipment, part of the Earthquake Arrival Recording Seismic System (EARSS) in New Zealand *(New Zealand © GNS Science/SSPL/ The Image)*

speed of sound is 660 miles per hour (MPH) (1,056 km/hr), compared to 760 MPH (1,216 km/hr) at the surface. (Temperature also affects the speed of sound.) In water, sound waves travel about five times faster than in air. In diamond, one of the hardest substances, sound travels about 40,000 MPH (64,000 km/hr)! Compression waves generally travel faster than shear waves in solids, since solids tend to be more difficult to compress than to bend or twist (which is what shear forces will do). Shear waves do not propagate in water because water does not resist shear forces.

Seismic or earthquake waves share these properties, and come in two varieties—compression waves and shear waves. (The term *seismic* derives from a Greek word, *seismos,* meaning shock or earthquake.) An earthquake is a violent movement of the earth as a result of built-up stresses that suddenly cause cracks to form and large masses of rock to move. (Chapter 6 discusses earthquakes in more detail.) This disturbance sends waves propagating out in all directions, just as a clap of a person's hands sends sound waves traveling in every direction. The seismic waves consist of motions of interior rock as well as rocks at the surface of the planet, along with soil and anything attached to the surface, such as buildings, roads, and bridges. Geologists record seismic waves with instruments called seismometers that detect motion in or along the ground as the waves pass.

Seismometers that are extremely sensitive can detect tremors from all over the globe, although the energy of a propagating wave dissipates, or dampens, as it travels because some of the motion is transformed into heat. Geologists from all over the world maintain an array of sensors to detect earthquake waves and to pinpoint the disturbance's origin, which is called the earthquake's *focus.* For instance, the *United States Geological Survey (USGS),* an agency devoted to Earth science and mapping, maintains a network of about 7,000 earthquake sensor systems in the United States. USGS is an extremely important contributor to geological research, as described in the following sidebar.

As the seismic waves spread out from the earthquake's focus, they travel at certain speeds. The fastest waves are the compression waves, which arrive at the sensor stations first and are called *P waves* or *primary waves.* P waves travel through rock at an average speed of about 13,000 MPH (20,800 km/hr) and through water and air at about the same speed as sound. *Secondary waves* or *S waves* are shear waves that propagate at a little more than half the speed of P waves. Because S waves are shear waves, they cannot propagate through liquids. Other types of waves are involved in earthquakes but are less important for studying Earth's interior.

In 1935 the California Institute of Technology researcher Charles Richter (1900–85) established a scale to measure the intensity of earthquakes. The *Richter scale,* which is still sometimes used, calculates the magnitude of an earthquake based on seismic wave amplitude—the

United States Geological Survey (USGS)

Land surveys to delineate boundaries and establish maps have always been an important function of governments. After the United States won its independence in the Revolutionary War, the government established a Surveyor General in 1796 and tasked this office with surveying western territories. Much of this land was sold or granted to the public, but the disposition of mineral lands—areas rich in natural resources—generated a lot of debate as to who got what and where. The science of geology was in its infancy at the time, so people had trouble determining where the natural resources were buried. But as the science grew and developed, geologists became more effective at locating resources, and on March 3, 1879, President Rutherford Hayes signed a bill establishing a new agency, the United States Geological Survey (USGS). The job of this agency was to classify lands according to their geological properties and mineral resources.

USGS's responsibilities have grown tremendously since its establishment. Although finding minerals and natural resources

size or extent of the vibrations. But the speed and type of the seismic waves, and where they are recorded, are more important for the study of the planet's interior.

INSIDE THE PLANET

Seismologists—geologists who study seismic waves—noticed in the early 20th century that P waves bended, or refracted, in their journey through Earth. Observations at stations far removed from the earthquake focus recorded waves that had traveled through the planet's interior, as illustrated in part (1) of the figure on page 12. Travel times of these waves indicated a refracted path, as shown in the figure, and wave remains a valuable service, geologists have expanded their knowledge and expertise into all aspects of Earth science, environmental issues, and biological phenomena. USGS employs 10,000 researchers and support staff to study and understand the planet and its resources, to reduce the danger and negative effects of natural disasters such as earthquakes and landslides, and to manage natural and environmental resources.

Among the agency's many projects are Priority Ecosystems Science, which supports the management of ecosystems that are of concern and value to society and is currently studying Florida's Everglades, San Francisco Bay, the Mojave Desert, the Platte River, and the Chesapeake Bay. USGS also maintains the Earthquake Hazards Program and the Advanced National Seismic System, which monitors about 20,000 earthquakes occurring in the United States each year. (Most are too small to be felt, but are important indicators of stress and *strain* at various locations.) Other programs involve energy resources, coastal and marine geology, habitats, water resources, fisheries, volcano hazards, and remote sensing with satellites.

speed is the distance divided by time (as determined by the amount of time elapsed since the start of the earthquake). Refraction was not too surprising because the increased pressure in Earth's interior results in firmer structures and more resistance to oscillation, so the wave speed is greater and seismic waves refract. What surprised early seismologists was that beyond a certain point—about 7,200 miles (11,600 km) from the focus, at an angular distance of 105 degrees—S waves disappeared!

In 1906 the British seismologist Richard D. Oldham (1858–1936) proposed that the disappearance of the shear waves was due to the "shadow" of a liquid core. Since S waves are shear, they cannot propagate through liquid, so the existence of a liquid center inside the planet would explain why seismometers fail to record shear waves on the other side of the planet from the focus, as shown in part (2) of the figure below. P waves, being compression waves, refract at the boundary between rock and liquid, creating a smaller "shadow." The rocky interior beneath the crust is called the *mantle,* and in 1914 the German seismologist Beno Gutenberg

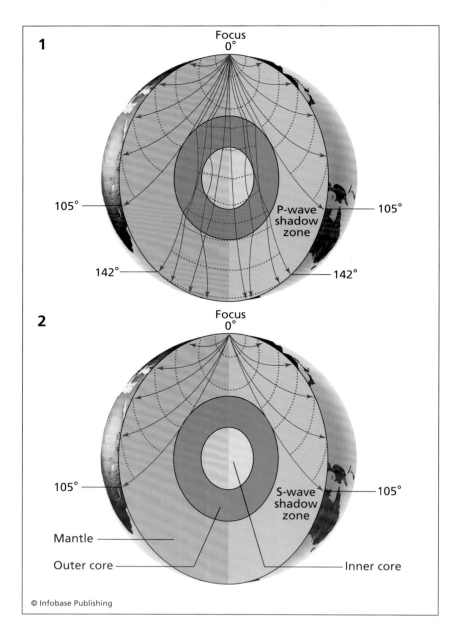

(1889–1960) used the seismic wave results to calculate that the mantle-core boundary is located at a depth of about 1,800 miles (2,900 km) below the surface.

However, in 1936 the Danish seismologist Inge Lehmann (1888–1993) analyzed seismic wave data and discovered an additional refractory step of P waves. Her analysis suggested the existence of another boundary, which she placed at a depth of about 3,200 miles (5,150 km). This boundary is between an *outer core* and an *inner core*.

The use of seismic waves to image Earth's interior is similar to the use of ultrasound waves to image the body's interior or sound waves in sonar to image the seafloor. Unlike ultrasound and sonar techniques, though, seismologists usually do not generate seismic waves—these are natural occurrences beyond the control of researchers. Yet the waves reveal a lot of information about otherwise inaccessible places. Seismic waves are also plentiful; about 1 million or so earthquakes occur each year in the world, and although most of these are fortunately minor they are detectable with sensitive instruments.

By studying the nature and speed of seismic waves, geologists have learned much about the Earth's interior. Earth consists of the following several layers:

- crust, composed of rocks having relatively low density, extending from the continental surface to an average depth of about 22 miles (35 km) and from the ocean floor an average of about four miles (6.4 km) down to a boundary known as the *Mohorovicic discontinuity* (*Moho* for short), named after the Croatian scientist Andrija Mohorovičić (1857–1936);
- mantle, extending from the crust to about 1,800 miles (2,900 km) below the surface, and divided into an upper and a lower section;
- outer core, which is liquid and extends from the mantle border to a depth of about 3,200 miles (5,150 km);

(opposite page) (1) Boundaries between the layers of Earth's interior bends or refracts P waves, causing shifts in speed and altered paths that leave "shadows"—areas that receive few or no waves.
(2) S waves fail to penetrate the liquid outer core, leaving a large shadow on the other side of the earthquake's origin.

- inner core, which is solid, with a radius of about 750 miles (1,220 km).

The mantle gets its name from Wiechert, who thought of it as a coat that covered the core (mantle derives from the German word, *mantel,* for "shell" or "coat"). About 67 percent of Earth's mass is contained in this large region. The mantle is mostly solid, although as discussed below there is some degree of fluidity in spots; it consists of minerals such as olivine and another silicate called perovskite ($MgSiO_3$). Silicon and aluminum are less abundant in the mantle compared to the crust, but magnesium is much more plentiful.

Wiechert assumed from the studies of Earth's density that the core must be dense. A greater density for the core also makes sense because the large portion of the heavier elements would have sunk to the interior as the hot, molten planet formed long ago. Iron and nickel possess relatively high densities and are commonly found in certain meteorites, indicating their abundance throughout the solar system. These metals are likely constituents of the core. The absence of shear wave propagation indicates the outer core is liquid, but studies of other seismic waves indicates a density slightly less than that expected if the outer core contained only melted iron and nickel. Instead, the outer core is about 90 percent iron and nickel, and most of this is iron—about 85 percent of the outer core is made of this element. The remaining 10 percent consists of lighter elements such as sulfur and oxygen.

The inner core forms a boundary with the outer core, reflecting some of the waves and transmitting the rest. Shear waves cannot pass through the outer core, but as compression waves cross the boundary between the inner and outer core, some of these disturbances create shear waves. The shear waves travel through the inner core and get converted back into compression waves as they proceed from the inner to the outer core. Seismologists can detect the paths of these waves, and the propagation of shear waves in the inner core implies it cannot be liquid. Density studies suggest the inner core is mostly solid iron, mixed with a small percentage of nickel.

Researchers continue to study seismic waves and similar data to learn more of the details on the structure and composition inside Earth. In 2005 John W. Hernlund and Paul J. Tackley of the University of California, Los Angeles, and Christine Thomas of the University of Liverpool in the Britain found data suggesting the presence of a thin layer around the mantle-core boundary. This layer, previously unknown and not yet widely studied, might help scientists to understand and identify further properties of the mantle. The researchers published their report "A Doubling of the Post-Perovskite Phase Boundary and Structure of the Earth's Lowermost Mantle" in a 2005 issue of *Nature.*

Although researchers can study the finer structure of Earth's hidden interior with sensitive seismometers, a large amount of information could also be gained by burrowing inside and taking a look. There are limitations on how far down people can drill, even with the hardest bits (the tip of the drill), but researchers are sharpening their drill bits in the effort to reach greater depths.

DRILLING INTO EARTH

Oil companies have drilled thousands of wells to extract subsurface oil. These wells range in depth from about 1,000 feet (305 m) to about 23,000 feet (7,000 m) and sometimes a little deeper. The deepest hole anyone has ever drilled as of 2009 is in Russia's Kola Peninsula, which is located in the northern part of the country, although the drillers were not searching for natural resources but instead were exploring how far down they could go. By the late 1980s, Russian scientists working in the Kola Peninsula reached a depth of 40,220 feet (12,262 m)—7.6 miles (12.26 km)!

Drilling to such depths is an extremely demanding operation. As the depth increases, the pressure increases and the rocks get harder, which results in slower progress and higher costs. Temperature also rises, as discussed in the following section, and the increased pressure and temperature greatly reduce the useful life of the expensive drill bits needed to cut through the hard earth (these drill bits cost $50,000 and sometimes even more). Controlling the drill and guiding its trajectory are not easy when the hole gets deep, and removing the cuttings from a great depth requires a lot of time and effort.

These difficulties make deep drilling a formidable task. But the difficulties have not stopped geologists from attempting ambitious projects. A U.S. project began in 1958 with the goal of drilling all the way to the Mohorovicic discontinuity, the boundary between crust and mantle. This project, called Project Mohole, would have been the first to reach

An oil rig platform off the California coast *(Chad Anderson/iStockphoto)*

the mantle, if it had been successful. Project Mohole failed to attain its primary goal, as discussed in the sidebar on page 18, due to budget problems and other daunting issues that the research team could not overcome.

Although Project Mohole failed to reach the mantle, a project with similar goals has recently emerged. Led by the Japan Agency for Marine-Earth Science and Technology (JAMSTEC), the project has been called Chikyu Hakken ("Earth discovery"). The primary objectives of this project are to observe and sample Earth's depths to obtain information about the nature and origin of earthquakes, as well as the structure and evolution of the planet. To achieve these ambitious goals, JAMSTEC ordered and received a vessel D/V *Chikyu* in July 2005. (D/V stands for drilling vessel.) Researchers and technicians outfitted the 689-foot (210-m) vessel with a drill system capable of drilling in 8,200 feet (2,500 m) of water and able to bore 23,000 feet (7,000 m)—4.3 miles (7 km)—into the seafloor. *Chikyu* cost about $550 million.

As part of the Integrated Ocean Drilling Program (IODP), supported by the United States and Japan with help from the European Union, China, and South Korea, *Chikyu* made its first expedition beginning in late 2007. In this first outing, researchers sailed to the Nankai Trough, an area of the Pacific Ocean off Japan's coast that has been the site of numerous earthquakes. Drilling in about 6,560 feet (2,000 m) of water, *Chikyu* cut a number of holes ranging in depth from 1,300 feet (400 m) to 4,600 feet (1,400 m) beneath the ocean floor. The sampled material proved to be relatively fresh as far as geology goes (4–6 million years) and appeared to be experiencing unusual amounts of stress.

Future *Chikyu* expeditions will drill even deeper holes. With its capacity to reach 4.3 miles (7 km) beneath the seabed, *Chikyu* should be able to achieve Project Mohole's goal of drilling into the mantle— the first time this layer will have ever been reached.

HEAT OF EARTH'S INTERIOR

One of the most interesting aspects of drilling into Earth is the rise in temperature with depth. This is not all that surprising to those people who have seen a volcano erupt and spew vast amounts of hot, molten rock called *lava*. The material comes from inside the planet, at places where hot, molten rock called *magma* has risen through cracks. (*Lava* is the term for this molten rock after the eruption; *magma* is generally the term used for subsurface molten rock.) Magma rises through these cracks because it is hotter and less dense than surrounding rock, similar to the way that hot air rises.

Visitors to Carlsbad Caverns, a group of caves in New Mexico, can descend about 830 feet (253 m) below the surface (some parts of the cave are deeper but not publicly accessible). Most visitors wear jackets because the temperature in these caves is about 56°F (13°C) all year. Although this temperature is cooler than the surface in summer months, the lack of sunlight and air movement results in a steady temperature. Geologists have measured Earth's temperature in mines with depths as great as 2.3 miles (3.78 km) and in smaller holes three or more times as deep, and these measurements show an average temperature increase of about 72°F/mile (25°C/km) in the crust, although the rate varies.

However, a temperature increase of 72°F/mile (25°C/km) cannot hold true throughout the mantle. At such a high rate, the lower regions of the mantle would be molten, but this is not consistent with seismic

Project Mohole—An Ambitious Attempt to Reach Earth's Mantle

Project Mohole was an attempt to drill a hole to the mantle and retrieve a sample from this great frontier—a frontier separated by vast quantities of hard rock. Suggested in 1957 by Walter Munk, a member of the U.S. National Academy of Sciences, the project got funds for preliminary work in 1958 from the National Science Foundation (NSF), one of the main government agencies that supports basic scientific research. A sample from the mantle would provide a large amount of information on the exact composition of this layer, its age, and internal dynamics. The question of mantle dynamics was particularly important during this time period, as continental drift was being hotly debated.

The thickness of Earth's crust varies widely, and the thinnest section is beneath the ocean. In some areas of the seafloor, the crust is only about three miles (4.8 km) thick, although the average is considerably more. The plan of Project Mohole consisted of three phases, the first of which was an experimental program to develop techniques to drill through deep water and into the crust. Drilling for oil in the relatively shallow areas of the sea is common, but Mohole scientists needed to drill in deeper parts of the oceans, in places where the crust is thinner. In the first phase of the project, beginning in early 1961, researchers drilled in 11,700 feet (3,570 m) of water off Guadalupe, Mexico. The platform was wave observations. The temperature gradient—change in temperature with depth—must be less in the mantle than in the upper part of the crust. Although the gradient cannot be measured directly, seismologists can make estimates based on seismic waves, taking advantage of the properties that depend on the nature of the rock through which the waves are traveling. For example, seismologists can determine the depth at which rocks begin to change phase, or state. Rocks change

a ship named *CUSS I,* a converted naval barge. (The ship's name came from the initial letters of the names of oil companies that had outfitted the ship—Continental, Union, Shell, and Superior.) Researchers drilled a series of holes, one of which extended into the ocean crust to a depth of 557 feet (170 m). Although this does not seem very far, the project became the first to drill successfully in deep water.

Phase two never got started. Cost estimates ballooned from $5 million to nearly $70 million. Although Phase one had succeeded, the project called for drilling through even deeper water and farther into the crust below, but no one was able to think of a cost-effective means of doing this. Project Mohole lost its funding in 1966 amid arguments about how the project should proceed and whether it was worth the money. (Another budget problem faced by Project Mohole was the existence of an even bigger and more expensive project that was competing for funds at the same time—the Apollo Moon landings.)

The project's failure was an embarrassment to the NSF, since the promising beginning had crumbled so quickly. A journalist Daniel S. Greenberg wrote a series of articles on the project in 1964 for *Science magazine,* and, as he watched the plan disintegrate, he wrote, "The Mohole business is a very sorry episode. . . ." Yet Project Mohole was not a complete failure, and geologists were able to identify a second sublayer of crust, consisting of rock called basalt, from the samples obtained at 557 feet (170 m) in the ocean crust.

A view inside Carlsbad Caverns near Devil's Spring *(Glenn Frank/iStockphoto)*

phase at certain temperatures and pressures, allowing geologists to calculate the temperature of these depths. Observations suggest that the mantle's temperature gradient is about 1.5°F/mile (0.5°C/km), much lower than the crust's.

Where does this heat come from? Earth's interior is hot for two main reasons. One source of heat is *radioactivity*—atoms of certain elements such as uranium and thorium undergo a natural process in which the atom's nucleus experiences a transformation, or decay, emitting energy in the form of radiation. Nuclear reactors use this same process to generate enough heat to turn huge turbines, producing large amounts of electricity. Radioactive atoms in Earth's interior are responsible for some of the heat inside the planet. The other source of heat is the remnants of energy created as bits of matter slammed into each other during Earth's creation. Although Earth formed billions of years ago, the violent collisions generated a lot of heat that remains trapped inside the planet.

Earth's core must be extremely hot. Unable to make a direct measurement, geologists can only estimate the core's temperature based on seismic wave calculations of pressure and composition. The temperature of the outer core probably exceeds 5,430°F (3,000°C). Even more uncertainty exists about the inner core's temperature, which may be as high as 14,400°F (8,000°C).

Hot objects cool off in three ways—radiation, *convection,* and conduction. Conduction carries away heat by contact with another object, such as the heat transfer that occurs when a person's finger comes into contact with a hot skillet. Convection involves currents such as air or liquid to carry away heat, such as the cooling effect of a sea breeze or fan. Radiation involves atomic emissions of electromagnetic energy in a frequency range that is commonly infrared—hot objects emit a lot of infrared radiation. Earth's surface radiates heat, which lowers the temperature (especially at night, when no sunlight is available to replenish it), but subsurface radiation does not escape. Heat from the interior flows through the depths by conduction and convection. The extent and mechanisms by which these processes occur are extremely important in understanding the structure of Earth's depths—and the movement of large chunks of crust and mantle.

TECTONIC PLATE MOVEMENT

Although Wegener's notion of continental drift was not entirely correct, researchers such as Harry Hess (1906–69) at Princeton University and Robert Dietz (1914–95) of Scripps Institution of Oceanography realized that Earth's crust separates at certain points in the middle of the ocean. At these sites, known as mid-ocean ridges, molten rock oozes upward to form a new seabed. A section of the Mid-Atlantic Ridge is shown in the figure. What causes the separation is the movement of rigid plates called *tectonic plates,* which were first postulated by the Canadian researcher J. Tuzo Wilson (1908–93) in 1965. The term *tectonic* derives from a Greek word, *tektonikos,* meaning "of a builder."

Earth's crust is composed of 12 large plates and a few dozen smaller ones. Plate boundaries do not necessarily follow continental boundaries, and the depth of the plates includes the crust plus a little bit of the upper part of the mantle. The crust and uppermost mantle composes

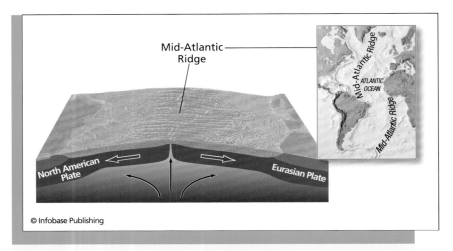

© Infobase Publishing

Two plates separate and move apart to form part of the Mid-Atlantic Ridge.

the *lithosphere* (from *lithos,* a Greek term for stone), which averages about 60 miles (100 km) in thickness. These rigid plates move around the surface and collide with other plates or move apart. A collision may send one plate buckling under the other, or the two plates may slide past one another. The motion is slow, in a range of 1–6 inches (2.5–15 cm) per year.

Plate movements have greatly affected the configuration of Earth's surface. At one time, millions of years ago, the seven continents were joined in one supercontinent known as Pangaea. (Named by Wegener, the term *Pangaea* is Greek for "all land.") The motion of the plates also helps explain earthquakes and volcanoes. For instance, a fissure or *fault* known as the San Andreas Fault in California lies around a boundary between two plates that grind past each other and periodically slip, causing earthquakes.

The forces at work to move the plates are of great interest to geologists. Plate motion requires some sort of flexibility in the layer of mantle on which the plates rest. This layer is known as the *asthenosphere* (from *asthenēs,* a Greek term for "weak"). Although the asthenosphere is not fully molten, it is not as rigid as the lithosphere, and is hot enough to deform or flow. An important component of this flow is a slow up-and-down circulation known as convection currents, which are driven by heat; hot material rises, cools as it loses heat to the surface, then falls back down, repeating the circulation when the deeper regions warm it up again.

Although geologists believe that motion from convection currents in the mantle drives the lithospheric plates, no one is certain exactly how this occurs or how far down the convection currents extend. A better understanding of these currents and their interaction with the plates would enhance geological knowledge on a variety of issues, including earthquakes and volcanoes. The discovery and modeling of new layers, such as the one found by Hernlund, Tackley, and Thomas, will help.

Careful monitoring of the plates reveals interesting plate motions that do not come directly from earthquakes—in other words, aseismic motions—the study of which may help explain the underlying processes. With global positioning system (GPS) equipment, which allows precision position measurements, geologists can detect subtle changes. With such sensitive instruments, Vladimir Kostoglodov of the National Autonomous University of Mexico and his colleagues detected a brief reversal in the motion of the plate at Guerrero, Mexico, that they cannot explain. The effect this strange motion may have on earthquake hazards in the area is unknown. Further research into the activity of Earth's interior is needed to clarify the issue.

DYNAMICS AND INTERACTIONS OF EARTH'S INTERIOR

Plate movements and mantle convection currents demonstrate how dynamic and changing Earth can be. Although these changes happen slowly, they produce significant effects, such as the rearrangement of the planet's surface.

Another important effect is the creation of Earth's strong *magnetic field.* A *magnet* has two *magnetic poles,* north and south, and Earth behaves in many ways as a gigantic magnet, with the north pole of the magnet somewhat close to the North Pole (which is located along the planet's rotational axis), and similarly for the south pole. This field aligns compass needles and deflects charged particles in space, creating spectacular displays of light such as aurora borealis (northern lights) and aurora australis (southern lights), as if there was a huge magnet embedded in the planet. But the cause of Earth's magnetic field is not a permanent magnet inside the planet; although the core is mostly iron, which is a highly magnetic material, the high temperatures of Earth's interior disrupt iron's magnetic properties, and the core is too hot to behave like an ordinary magnet. As described in chapter 2, geologists believe that convection currents in the iron core generate Earth's magnetic field. The mechanism that produces the field is sometimes called a geodynamo.

Interactions also play a role in the properties and behavior of Earth's interior. The boundaries between layers are crucial in transmitting or reflecting seismic waves and serve as the sites where two different materials come into contact and interact. For example, the liquid outer core, rich in metals, and the silicate rock of the deepest mantle meet at a depth of about 3,200 miles (5,150 km).

The great depth of regions, such as the mantle-core boundary, makes these areas impossible to sample directly. Yet geologists are developing other means to study possible interactions.

Leslie A. Hayden and E. Bruce Watson, researchers at Rensselaer Polytechnic Institute in Troy, New York, have found a mechanism by which metal atoms from the core can leak, or diffuse, across the boundary. These researchers studied the mantle-core boundary by creating an artificial boundary in the laboratory. They constructed a silicate material having a composition similar to what geologists believe is in the mantle and placed it next to metallic material. Then the researchers heated and pressurized the materials to reproduce conditions in Earth's interior at the depth of the mantle-core boundary. Hard rock, especially under high pressure, would seem to offer few if any avenues for metals to enter, yet Hayden and Watson discovered metal atoms crossed the boundary. These metals included elements that exist in small quantities in the core, such as gold and platinum. What causes the atoms to move across the boundary is not clear, but the researchers propose the atoms diffuse between crystals, or grains, of the rock. Hayden and Watson published their findings, "A Diffusion Mechanism for Core-Mantle Interaction," in a 2007 issue of *Nature*. Such interactions may play a vital role in the distribution of elements and the chemical composition of Earth's interior.

CHARTING THE DEPTHS WITH RESEARCH IN THE LABORATORY

The experiments of Hayden and Watson illustrate the use of experimental techniques to study phenomena hidden far below the surface of the planet. Equipment to generate high temperatures and pressures that mimic Earth's interior has allowed geologists to bring some of their studies into the laboratory. One of the most common laboratory tools is the diamond anvil cell.

Diamonds are the hardest natural material, which makes them excellent components for a cell, or container, in which high pressure is to be generated. An anvil is a block capable of withstanding high pressures or hammering, such as the steel anvil on which metalworkers once hammered and molded swords and other objects. In a diamond anvil cell, two blocks made of diamond press against the material to be studied, squeezing it and exerting tremendous pressure.

Considering the high cost of diamonds and other sufficiently hard substances, these anvil cells are not usually very large. As a result, most laboratories can subject only a small amount of material to high pressures in any given experiment. Maintaining a high temperature is also a problem, since heat readily flows out of the anvils, and the high temperatures can weaken the diamonds by loosening their structure. Yet these cells can exert a pressure in excess of 1 million times as strong as the atmosphere—comparable to the pressure at Earth's center.

Geologists who use diamond anvil cells and similar equipment can study the properties that rocks have under the extreme conditions of Earth's interior. For example, Jonathan C. Crowhurst of the Lawrence Livermore National Laboratory in California, along with colleagues at the University of Washington, Carnegie Institution of Washington in Washington, D.C., and Northwestern University in Illinois, studied a mineral known as ferropericlase. This mineral, which consists of magnesium (Mg), iron (Fe), and oxygen (O), is common in the lower depths of the mantle. (Although no one has sampled the mantle directly, the study of seismic waves and the analysis of material such as magma and diamonds that have risen from the depths have given geologists some idea of mantle composition.)

Crowhurst and his colleagues applied pressures of up to about 600,000 times that of Earth's atmosphere to ferropericlase and then measured a property known as spin transition. This property has an important effect on elasticity—how readily the molecules of a substance move around—which influences the conduction of seismic waves and is critical for the study of the mantle. As the authors wrote in their report, "Elasticity of (Mg,Fe)O Through the Spin Transition of Iron in the Lower Mantle," in a 2008 issue of *Science*, "Because knowledge of this deep and inaccessible region is derived largely from seismic data, it is essential to determine the influence of the spin transition on elastic wave velocities at lower-mantle pressures."

Many materials change properties at high pressure and temperature. But Crowhurst and his colleagues discovered that ferropericlase experienced more changes than had been expected, causing the speed of seismic waves to slow down a little bit. This finding is important to seismologists, who must take these factors into account during the analysis of seismic wave data.

Advances in computers have also created valuable opportunities for geologists. The fastest computers, known as supercomputers, perform trillions of operations per second. Geologists simulate the physical and chemical properties of matter with sophisticated computer software, including programs that incorporate mathematical equations describing these properties and the interactions of matter at extremely high temperature and pressure. Simulations always rely on the accuracy of scientific knowledge—if the properties and interactions incorporated into the computer program are wrong, the results will also be wrong. But if geologists are careful to use the findings of previous experiments, such as laboratory experiments using diamond anvil cells, a computer simulation is a useful tool. A computer simulation lets geologists explore down to the atomic level what may be happening all the way inside Earth's core.

Anatoly B. Belonoshko at the Royal Institute of Technology in Stockholm, Sweden, and his colleagues simulated iron atoms under the conditions the atoms experience in the inner core. When in the solid phase, iron atoms adopt a certain geometric configuration, as do many other atoms. This configuration forms a repeating structure called a crystal. Belonoshko and his colleagues conducted computer simulations of iron to indicate what sort of crystal structure may exist in Earth's inner core.

One of the reasons crystal structure is important is that it will influence elasticity and therefore seismic wave conduction. Seismologists have determined that the inner core shows elastic anisotropy, which means that its elastic properties depend on direction. For example, seismic waves travel faster when they are moving in the same direction as Earth's axis than when they are moving perpendicular to this direction.

What causes this anisotropy? One possible explanation is that the iron crystals composing the core have a particular orientation, so that waves traveling along this direction would have a different speed than waves traveling, say, perpendicular to it. But iron tends to become isotropic—without orientation—at high temperature and pressure.

As an alternative hypothesis, Belonoshko and his colleagues suggested that iron in the core adopts a certain crystal pattern called body-centered cubic, in which the atoms form a cube with an atom in the middle. The researchers conducted simulations using a method called molecular dynamics, which incorporates atomic interactions. In their report, "Elastic Anisotropy of Earth's Inner Core," published in a 2008 issue of *Science,* Belonoshko and his colleagues wrote, "We show, by molecular dynamics simulations, that the body-centered cubic iron phase is extremely anisotropic to sound waves despite its high symmetry. Direct simulations of seismic wave propagation reveal an anisotropy of 12 percent, a value adequate to explain the anisotropy of the inner core." These simulations suggest that the core's anisotropy is not due to a particular orientation of the iron but to the crystal itself.

CONCLUSION

Geologists will continue to complement field studies and seismic wave observations with laboratory experiments and computer simulations. Advanced technologies such as the drilling vessel *Chikyu* create opportunities for researchers to explore previously unreachable depths, and the samples obtained from these operations will enhance knowledge of the structure and composition of the upper mantle. But the extreme pressure and temperature of Earth's interior probably set limits on how far down people can ever drill. Although these limits constrain scientists' reach, simulations and laboratory experiments, coupled with seismology, extend all the way to the center of the planet.

Another approach to the study of Earth's depths has recently gained interest. This approach involves extending geology's reach not only to Earth's center but also throughout the entire solar system. The solar system evolved and gave birth to the Sun and the planets about 4.5 billion years ago. Astronomical and geological evidence provide clues about this event, which involves an enormous, swirling cloud of dust and gas that eventually aggregated into the Sun and planets. Yet the details are not at all clear. Studying the birth and evolution of the solar system will help scientists understand how the system's bodies formed, which would also help explain their present structure.

For example, Earth and the planet Mars have many similarities. Mars is smaller, having a radius a little more than half that of Earth, and has a density of about 73 percent the value of Earth's density. Its orbit is about 1.5 times larger than Earth's orbit, which places it about 45 million miles (72 million km) farther away from the Sun on average. Probes launched in the United States and other nations have reached Mars, orbiting the planet and in some cases landing on its surface, mapping the terrain and analyzing soil chemistry. Although seismic data from Mars is not yet available, density and gravity measurements of Mars suggest it has a core similar to Earth's, although perhaps containing a higher percentage of lighter elements.

No probes or spacecraft have yet been sent to retrieve samples from Mars, but scientists have a rare but valuable opportunity to study material from this planet. Meteorites—rocks from space—sometimes land on Earth. Many of these meteorites come from leftover debris from the solar system's formation, but some of these rocks display chemical compositions indicating that they came from Mars. (Violent collisions or other activity ejected these rocks from the surface of Mars with sufficient speed to escape the planet's gravity.) Only a few dozen of the thousands of meteorites found on Earth are from Mars, but these rocks provide insights as well as informative comparisons with Earth. In 1996 a team of researchers at the National Aeronautics and Space Administration (NASA) announced that they had found fossils in one of these

Martian meteorites—which indicates life evolved on Mars—but their results are controversial.

Alex N. Halliday and R. Bastian Georg of the University of Oxford in Britain, along with colleagues at the University of California, Los Angeles, and the Swiss Federal Institute of Technology Zurich, recently studied a variety of meteorites. The researchers focused on silicon, the second most common element in Earth's crust and an abundant element throughout the solar system. Silicon, like other elements, has different *isotopes*—atoms that have the same number of protons (which specifies the element) but a varying number of neutrons. Although isotopes tend to have similar chemical properties, they possess different masses, which gives them slightly different physical properties. When Halliday, Georg, and their colleagues compared silicon isotopes in Earth and Mars material, they found that Earth silicates have a greater proportion of heavier isotopes (isotopes with more neutrons). This evidence suggests that Earth and Mars may have formed under different conditions and may have distinctly different cores. The researchers published their report, "Silicon in the Earth's Core," in a 2007 issue of *Nature*.

Further studies of astronomical material, perhaps including samples retrieved from future space missions, will enhance knowledge of the solar system and all of its planets, including Earth. As science reaches out across the vast distances of space, scientists are also probing deeper into the very heart of the planet. Exploring Earth's depths is a science whose frontiers range from the great heat and pressure of the planet's core to the space probes that travel among the planets.

CHRONOLOGY

1875   The Italian researcher Filippo Cecchi (1822–87) builds one of the first seismometers, although the instrument is not very sensitive.

1879   The U.S. government establishes the USGS.

1896   The German scientist Emil Wiechert (1861–1928) hypothesizes that Earth contains a metal core surrounded by a rocky mantle.

1897   Wiechert improves upon seismometer technology, building an instrument that can record throughout an earthquake episode.

1906   The British seismologist Richard D. Oldham (1858–1936) analyzes seismic waves to show that part of Earth's core is liquid.

1909   The Croatian researcher Andrija Mohorovičić (1857–1936) analyzes seismic waves and finds the Mohorovicic discontinuity, which separates Earth's crust and mantle.

1912   The German researcher Alfred Wegener (1880–1930) proposes that Earth's continents drift over time.

1914   The German seismologist Beno Gutenberg (1889–1960) uses seismic waves to locate the depth of the mantle-core boundary at about 1,800 miles (2,900 km) below the surface.

1936   The Danish seismologist Inge Lehmann (1888–1993) analyzes seismic waves and discovers evidence for a boundary between a solid (inner) and liquid (outer) core, which she places at a depth of about 3,200 miles (5,150 km).

1958   The Project Mohole, an attempt to drill into the Mohorovicic discontinuity, begins. The project would last eight years but fail to attain its primary goal.

1965   The Canadian researcher J. Tuzo Wilson (1908–93) proposes the theory of plate tectonics.

1980s   The Russian scientists drilling in the Kola Peninsula reach a depth of 7.6 miles (12.26 km), the deepest hole ever drilled.

2005   The Japan Agency for Marine-Earth Science and Technology (JAMSTEC) begins testing the drilling vessel *Chikyu,* capable of drilling 4.3 miles (7 km) into the ocean floor.

2007   In an expedition to Nankai Trough, an area of the Pacific Ocean off Japan's coast that has been the site of numerous earthquakes and tsunamis, scientists aboard *Chikyu* drill holes ranging from 1,300 feet (400 m) to 4,600 feet (1,400 m) into the seabed.

FURTHER RESOURCES
Print and Internet

Belonoshko, Anatoly B., Natalia V. Skorodumova, Anders Rosengren, and Börje Johansson. "Elastic Anisotropy of Earth's Inner Core." *Science* 319 (February 8, 2008): 797–800. Belonoshko and colleagues suggest that iron in the core adopts a certain crystal pattern called body-centered cubic, in which the atoms form a cube with an atom in the middle.

Bjornerud, Marcia. *Reading the Rocks: The Autobiography of the Earth.* New York: Basic Books, 2006. Bjornerud, a geologist, chronicles the history of Earth as revealed by the rocks and layers that compose it. Starting at the very beginning, at the birth of the solar system, she discusses evolution, plate tectonics, climate change, and many other topics.

Brush, Stephen G. *Nebulous Earth.* Cambridge: Cambridge University Press, 1996. Suitable for advanced readers, this book details the fascinating work of the scientists who developed the concepts and principles of planetary geology and the evolution of the solar system.

Crowhurst, J. C., J. M. Brown, A. F. Goncharov, and S. D. Jacobsen. "Elasticity of (Mg,Fe)O Through the Spin Transition of Iron in the Lower Mantle." *Science* 319 (January 25, 2008): 451–453. Crowhurst and his colleagues discovered that the properties of certain materials result in a slowing of the speed of seismic waves.

Dixon, Dougal. *The Practical Geologist: The Introductory Guide to the Basics of Geology and to Collecting and Identifying Rocks.* New York: Simon and Schuster, 1992. This book introduces the subject of geology and focuses on practical applications, such as collecting minerals and making maps.

Georg, R. Bastian, Alex N. Halliday, Edwin A. Schauble, and Ben C. Reynolds. "Silicon in the Earth's Core." *Nature* 447 (June 28, 2007): 1,102–1,106. This research indicates that Earth and Mars may have formed under different conditions and may have distinctly different cores.

Greenberg, D. S. "Mohole: The Project That Went Awry." *Science* 143 (January 10, 1964): 115–119. The sad history of Project Mohole is chronicled here.

Hayden, Leslie A., and E. Bruce Watson. "A Diffusion Mechanism for Core-Mantle Interaction." *Nature* 450 (November 29, 2007): 709–711. These researchers have found a mechanism by which metal atoms in Earth's core can leak, or diffuse, across layer boundaries.

Hernland, John W., Christine Thomas, and Paul J. Tackley. "A Doubling of the Post-Perovskite Phase Boundary and Structure of the Earth's Lowermost Mantle." *Nature* 434 (April 14, 2005): 882–886. This paper describes data suggesting the presence of a thin layer around the mantle-core boundary.

Japan Agency for Marine-Earth Science and Technology (JAMSTEC). "*Chikyu Hakken.*" Available online. URL: http://www.jamstec.go.jp/chikyu/eng/index.html. Accessed May 4, 2009. The English version of JAMSTEC's Web pages on their Earth Discovery project contains information on the drilling vessel *Chikyu* and its expeditions, along with the latest findings.

Louie, John N. "Earth's Interior." Available online. URL: http://www.seismo.unr.edu/ftp/pub/louie/class/100/interior.html. Accessed May 4, 2009. Beautifully illustrated, this essay discusses the structure of the planet and how geologists discovered this structure.

Mathez, Edmond A., ed. *Earth: Inside and Out.* New York: New Press, 2001. Written by a team of experts, this highly informative book contains sections on Earth's evolution, seismic exploration of the interior, plate tectonics, analysis of rocks, and climate change.

ScienceDaily. "2006 Tectonic Plate Motion Reversal Near Acapulco Puzzles Earthquake Scientists." News release, August 6, 2007. Available online. URL: http://www.sciencedaily.com/releases/2007/08/07080213

0847.htm. Accessed May 4, 2009. Vladimir Kostoglodov of the National Autonomous University of Mexico and his colleagues spotted an unusual reversal in the motion of the plate at Guerrero, Mexico.

———. "Deep-Sea Drilling Yields Clues to Mega-Earthquakes." News release, December 18, 2007. Available online. URL: http://www.sciencedaily.com/releases/2007/12/071212201948.htm. Accessed May 4, 2009. A description of the findings of an expedition of the scientific drilling vessel *Chikyu* to the Nankai Trough.

University of California Museum of Paleontology. "Plate Tectonics." Available online. URL: http://www.ucmp.berkeley.edu/geology/tectonics.html. Accessed May 4, 2009. Part of an online exhibit, this Web resource includes links to essays on the history and mechanisms of plate tectonics, along with movies and animations that illustrate the basic concepts.

UPSeis. "What Is Seismology and What Are Seismic Waves." Available online. URL: http://www.geo.mtu.edu/UPSeis/waves.html. Accessed May 4, 2009. UPSeis is an educational site aimed at young people interested in seismology. This Web page explains the nature of seismic waves and includes several helpful diagrams.

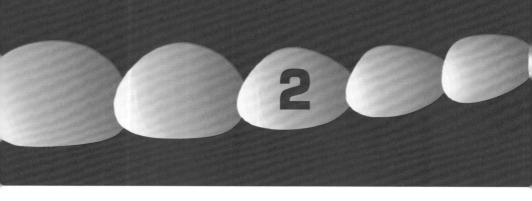

ORIGIN AND VARIABILITY OF EARTH'S MAGNETIC FIELD

About 1,000 years ago, people in China began using iron or an iron-bearing mineral called magnetite (or lodestone) as a direction finder. When free to rotate, a needle made of magnetite, or iron rubbed with magnetite, aligns itself in a north-south direction. This directional effect, due to magnetism, became the basis for the compass. The details of the discovery and origin of the compass are lost in the veil of time, but by the 13th century compasses were playing critical roles in navigation and trade in many parts of the world. Sailors crossing the open sea used the Sun and stars for guidance, but the sky was sometimes cloudy, and interpreting the movements of astronomical bodies often depends on the time of day, the season of the year, and the sailor's position. Compasses are simple and reliable.

A theory of how compasses work did not come until centuries later. William Gilbert (1544–1603), a British physicist and physician, studied compasses and magnetism in the late 16th century. In 1600 Gilbert published *De Magnete* (Latin for "On the Magnet"), a book in which he recorded his findings and proposed a theory. Compasses point northward, Gilbert claimed, because Earth is a gigantic magnet that exerts a force. Magnets are usually made of iron or iron-bearing minerals, which are commonly found on Earth's surface or under the ground. The magnetic effect that Gilbert hypothesized for Earth is similar to a bar magnet acting on iron filings, aligning the little bits of iron to its lines of force. The lines of force are associated with a magnetic field—a region of space in which magnetic forces act. Earth's magnetic field is also known as the geomagnetic field (*geo* is a Greek prefix meaning Earth). According to Gilbert, compasses align themselves to the geomagnetic field.

Gilbert's ideas seemed to explain the behavior of compasses. Yet navigators began noticing that Earth's magnetic field ·was not constant. Instead of always pointing in exactly the same direction, compasses deviated, changing direction slightly over the years. These shifts were difficult to understand if Earth was a fixed bar magnet. The origin and nature of Earth's magnetic field appeared to be more complicated.

Geologists study Earth's magnetic field because it is critical for many applications—although global positioning system (GPS) receivers have largely replaced compasses for navigation these days, Earth's magnetic field influences radio communication and other important technologies. Earth's magnetic field also reveals much about the structure of the planet. The previous chapter described Earth's core, which is mostly made of iron. Earth's core is the basis for the planet's magnetic field, but the mechanism is not as simple as Gilbert envisioned. This chapter explains how and why scientists have reached this conclusion. Although researchers have made progress in understanding the complicated phenomena underlying Earth's magnetic field, much crucial information remains undiscovered at this frontier of Earth science.

INTRODUCTION

Magnetism is closely related to electricity, although this relationship is not obvious and took many years for scientists to appreciate. In 1820 the Danish physicist Hans Christian Oersted (1777–1851) found that an electric *current* produces a magnetic field. A current is a flow of electric charges, and when charges flow along a conductor such as a wire, the conductor creates a magnetic field. Oersted measured this magnetic field by the force it exerted on a compass needle in its vicinity. In the 1830s the British scientist Michael Faraday (1791–1867) discovered a similar but opposite relation—a changing magnetic field induces an electric current in a conductor. The Scottish physicist James Clerk Maxwell (1831–79) formulated a set of equations in the 1860s describing the

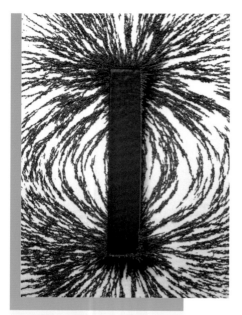

Iron filings align themselves to a bar magnet's field, showing the lines of force. *(Cordelia Molloy/Photo Researchers, Inc.)*

mathematical behavior of electric and magnetic fields. Maxwell showed that these fields arise from interactions of electrically charged particles—these interactions, and the associated forces, are known as electromagnetism.

A magnet exerts a force on other magnets, although the nature of the force depends on the magnets and their orientation. Common magnets such as a bar magnet are dipoles, meaning they have two magnetic poles or ends, one of which is called north and the other south. (These terms reflect the importance of compasses in the early studies of magnetism.) As one magnet approaches another, the north pole of each magnet attracts the south pole of the other magnet, while the north pole repels the north pole of the other. The same is true for south poles, which attract the north pole of another magnet but repel the south pole. Magnets also tend to affect metallic objects in their vicinity, especially ones containing iron, even if those objects do not appear to be strongly magnetic.

What gives a magnet its magnetic properties? Notice that there are several types of magnets. One type, sometimes called a permanent magnet, is made of iron, such as a bar magnet. The other type of magnet is an electromagnet; as Oersted discovered, an electric current exerts a magnetic force, and an electromagnet is a conductor capable of carrying a current. Electromagnets have the advantage of being easily switched off, which the operator can do by cutting off the current.

All magnets and magnetic forces involve electric charges, as Maxwell deduced, although the electrical contribution is more obvious in electromagnets. Iron magnets derive their properties from ferromagnetism. (*Ferrum* is a Latin word meaning iron.) Ferromagnetism is not limited to iron, but it is a property that is especially prominent in materials containing iron, nickel, or cobalt. A ferromagnetic material can become magnetized—exert magnetic forces—if it is exposed to a strong magnetic field. And it will remain magnetized after the strong field is removed. A full explanation of ferromagnetism involves advanced concepts in physics such as quantum mechanics; a brief explanation is that interactions between atoms, and certain properties of negatively charged *electrons* orbiting an atom's nucleus, are responsible.

Electrons are generally constituents of all atoms, but what makes ferromagnetic materials special is that the atoms in these materials form special areas called domains. Exposure to an external (outside) magnetic field aligns the atoms in a ferromagnetic material, forming magnetic domains that line up or orient themselves in a similar direction. The combined effect is to strengthen the magnetic properties. If the external field is strong, domains in ferromagnetic substances remain oriented after the external field is gone, so the magnetization of the object remains—a magnet has been created. Iron and other ferromagnetic materials are also responsive to weaker magnetic fields—for example, even a weak magnet attracts iron filings—even though the field may not be strong enough to cause a permanent change.

But a "permanent" magnet is not necessarily permanent. If the magnetic domains are scrambled again, a magnet will lose its magnetism. Sometimes tapping or pounding a magnet is enough to destroy the domain alignment, but one of the most effective methods of demagnetization is to apply heat. Atoms and molecules are always in motion, even in a solid, at a speed that depends on temperature, and higher temperatures elevate the average speed. When agitated by heat, atoms and molecules may move around so much that the domains jiggle out of alignment. Above a certain temperature called the Curie temperature (after the French scientist Pierre Curie [1859–1906], the husband of Marie Curie [1867–1934]), a ferromagnetic material loses its "permanent" magnetism. The Curie temperature varies for different materials; iron's Curie temperature is 1,418°F (770°C).

Earth's magnetic field resembles the field of a dipole magnet. As an approximation, Earth behaves similarly to a bar magnet, as Gilbert proposed, but the bar is not aligned with the planet's axis of rotation, as illustrated in the following figure. This means that the north pole of the magnet is not located at the same point as the North Pole, which is directly on Earth's axis, but instead is a small distance away. A compass

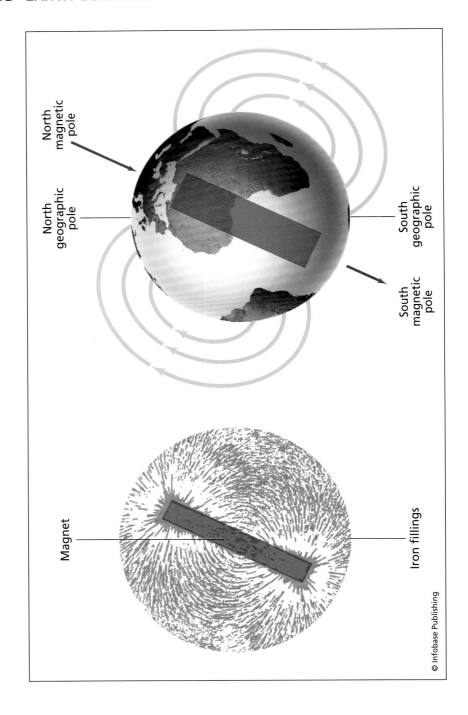

(opposite page) Earth's magnetic field behaves approximately as if it were coming from a bar magnet buried in the planet, although the magnetic poles are at a slight angle (roughly 11 degrees) from the axis of rotation (North and South Poles).

does not point directly north but rather indicates the direction of the north magnetic pole.

EARTH'S MAGNETIC FIELD

In most places outside of the polar regions, compasses work well, indicating the north/south direction. People are not the sole users of Earth's magnetic field for this purpose. Migratory birds appear to use the geomagnetic field to help them navigate the long journeys they make to escape harsh winter environments. Other animals also seem to have a magnetic sense. This magnetic sense may consist of an internal compass, composed of tiny crystals of magnetite, or it may be due to chemical or electrical effects of magnetic fields.

Although Earth's magnetic field is useful, Earth is not simply a large bar magnet. Chapter 1 discussed how geologists explore the interior of the planet, including its large metallic core. The core is composed of a solid inner core having a radius of about 750 miles (1,200 km) and consisting mostly of iron. Surrounding this inner core is a liquid outer core that extends another 1,400 miles (2,250 km), so that the radius of the entire core is about 2,150 miles (3,450 km). The outer portion is mostly iron and nickel, along with a small percentage of lighter elements. Both inner and outer portions are remarkably hot—the outer core's temperature is probably at least 5,430°F (3,000°C), and the inner core may be as high as 14,400°F (8,000°C). These temperatures are far above the Curie temperature at which iron loses its ferromagnetic properties.

To study Earth's magnetic field, geologists map its characteristics from the surface. During an expedition to the extreme northern latitudes of the Arctic, the British explorer Sir James Clark Ross (1800–62) located the north magnetic pole in 1831. As can be seen in the figure on

Geological Survey of Canada

The GSC was created in 1842. A year earlier, the legislature of the Province of Canada, which at the time consisted of parts of modern day Ontario and Quebec, resolved to fund a geological survey of the province, and the agency born to carry out this survey became the GSC. The motivation for establishing GSC was similar to that which was to lead to the creation of the USGS in Canada's southern neighbor in 1879—to assess the natural resources of the land. Canada possesses considerable natural resources and is among the world's leading producers of copper, zinc, nickel, uranium, and other minerals. Oil and natural gas deposits in and around the country are also rich; in 2006, for example, Canada was the leading exporter of crude oil to the United States, accounting for about 20 percent of the total U.S. imports for this crucial energy resource, according to official statistics of the U.S. government.

Surveys of Canada are daunting for several reasons. Canada has a total area of approximately 3,855,000 miles^2

page 38, the lines of force at the north and south magnetic pole dip into the ground. A sensitive compass will dip at a 90 degree angle, pointing straight down, at the north magnetic pole. In 1831, this point was on the west coast of the Boothia Peninsula in northern Canada, in what is now Nunavut. The Norwegian explorer Roald Amundsen (1872–1928) became the next person to find the northern magnetic pole, which he accomplished in 1904 during an Arctic expedition. But the north magnetic had moved about 30 miles (50 km) north since Ross's time!

Canadian geologists have tracked the movement of the north magnetic pole since 1948. In 1948 the Canadian researchers Paul Serson and Jack Clark located the 90 degree dip near Allen Lake on Prince of Wales Island, about 155 miles (250 km) northwest of Amundsen's discovery.

(10,000,000 km²), making it the second largest country in the world behind Russia. About 10 percent of Canada's surface area is freshwater. With 150,000 miles (240,000 km) of shoreline, Canada has more shoreline than any other country in the world. The climate in most of the northern portion of the country is cold, harsh, and challenging.

As part of Canada's Earth Sciences Sector, GSC will continue to aid the development of the country's rich natural resources as well as conduct other geological projects. In addition to observing the position of the north magnetic pole, GSC research includes environmental studies, monitoring hazards such as earthquakes and landslides, and glaciology (the study of ice and glaciers). For example, much of the northern areas of the country consists of permafrost, defined as soil or rock that is frozen for much of the year. Permafrost thickness depends on the properties of the soil and its vegetation, along with local temperature and climate. The study of permafrost, along with other geological features that are sensitive to the climate, will contribute to the ongoing worldwide research efforts to study global climate change.

Since then, periodic expeditions have shown that the north magnetic pole has moved at an average speed of about six miles (10 km) per year northward, and seems to be accelerating. The last expedition, which the Geological Survey of Canada (GSC) conducted in 2001, located the north magnetic pole at a latitude of 81.3°N, in the Arctic Ocean about 620 miles (1,000 km) from the North Pole. GSC is a geological information and research agency in Canada that is similar to the United States Geological Survey (USGS), which was discussed in a sidebar on page 10 in the previous chapter. As described in the sidebar on page 40, GSC has a long history of geological service and will continue to track the north magnetic pole and perform other important observations and research projects.

Trans-Antarctic Mountains in Antarctica *(Ardo X. Meyer/NOAA)*

If Earth was a huge bar magnet, the south magnetic pole might be expected at a location exactly opposite the north magnetic pole—the southern pole would be the other end of the bar. But the region at which Earth's magnetic field points upward is presently about 1,770 miles (2,850 km) from the South Pole, at a latitude of 64.5°S, near the coast of Antarctica. This location is not along a straight line through Earth connecting to the north magnetic pole. The south magnetic pole wanders in the same way as the north magnetic pole, varying independently though with a similar rate. Researchers at the Australian Antarctic Division and other organizations monitor the location of the south magnetic pole.

The movement of the magnetic poles suggests a dynamic process. Earth is a dynamic planet, with sliding tectonic plates, mantle convection, and a liquid outer core. Theories of the origin of Earth's magnetic field will be described in the section "Dynamo Theory of Earth's Magnetic Field" on page 47.

Another important question concerns the age of Earth's magnetic field—when did it form? Geologists try to answer this question by studying ancient rocks. Although the age of a rock is sometimes difficult to determine, geologists gather clues from the position and composition of the rock, as well as from radioactive atoms. The nucleus of these atoms decay at a constant rate, providing geologists with the means of gauging when the rock formed. During rock formation, such as when molten rocks cool and solidify or when sediments get compacted into sedimentary rock, the presence of a magnetic field can affect the rock's orientation and structure, particularly if it contains even a small amount of ferromagnetic substances such as iron. For instance, magnetic fields align iron-bearing crystals, giving most of these crystals a specific orientation, rather than having random orientations.

Some of the oldest known rocks, dating back 2 to 3 billion years, show evidence of exposure to a magnetic field. This finding suggests that Earth's magnetic field formed early in its history. But measuring the strength or intensity of this field is a much more elaborate procedure. Recently, John A. Tarduno, a researcher at the University of Rochester, and his colleagues analyzed samples of 3.2-billion-year-old rocks. Tarduno and his colleagues performed a difficult measurement of the magnetism in the rocks by heating tiny crystals and observing the magnetic fields with an extremely sensitive piece of equipment called a *superconducting quantum interference device (SQUID)*. The results, as reported in "Geomagnetic Field Strength 3.2 Billion Years Ago Recorded by Single Silicate Crystals," published in a 2007 issue of *Nature,* indicate Earth's magnetic field was about half as strong as it is today. As quoted in a ScienceDaily news release dated April 5, 2007, Tarduno commented, "These values suggest the field was surprisingly strong and robust. It's interesting because it could mean the Earth already had a solid iron inner core 3.2 billion years ago, which is at the very limit of what theoretical models of the Earth's formation could predict."

The strength or intensity of Earth's magnetic field is critical for its role in protecting the planet from high-speed particles emitted by the Sun. Earth's magnetic field provides this protection because the field extends from the planet out into space—the *magnetosphere*—and interacts with charged particles.

MAGNETOSPHERE

Magnetic fields exert a force on electric charges in motion—this is another aspect of electromagnetism and the close relationship between electricity and magnetism. The force acts perpendicular (at a 90 degree

Northern and Southern Lights

Aurora was the Roman goddess of dawn, and the term *bore-al* derives from a Latin word referring to the north. Because the northern lights often appear as if a sun was rising in the north, the phenomenon is called aurora borealis. The term for the southern lights is *aurora australis, australis* being Latin for southern.

Northern and southern lights occur most often in polar regions, within about 1,500 miles (2,400 km) of the magnetic pole. The displays usually last for only a few minutes (though some endure for hours), can be as bright as moonlight, and exhibit colors, most commonly green but also red, violet, and blue. In ancient times, people who lived in extreme northern latitudes explained the strange displays with myths and legends. Some legends in Finland ascribe the lights to a fable involving a fiery arctic fox, which accounts for the Finnish name for the northern lights, *revontulet,* meaning fox fires. Other peoples, including Vikings, considered the lights to be ghostly maidens.

The scientific cause of the northern and southern lights involves the magnetosphere, as suggested by the proximity of the light displays to the magnetic poles. Charged particles of the solar wind become accelerated as they interact with the magnetosphere. These interactions are complicated and not fully understood, but the result is an impulse directing charges such as electrons speeding along the magnetic field lines, which converge at the magnetic poles (see figure on page 38). Violent collisions between these charged particles and oxygen or nitrogen in the atmosphere cause the atmo-

angle) to the direction of the charge's motion. (If the electric charge is not moving, it experiences no force from the magnetic field.) This means that an electric charge such as an electron or proton moving

Aurora borealis in Anchorage, Alaska, in 1977 *(Yohsuke Kamide, Nagoya University/Collection of Herbert Kroehl/NGDC, NOAA, NWS)*

spheric gases to gain energy, which they may release in a burst of light.

Sometimes the light displays are especially intense. These occasions coincide with disturbances in the Sun that eject a greater than usual number of particles. For instance, in the summer of 1859, the Sun released a huge quantity of energy and particles in an event known as a solar flare. The intensity of this surge overwhelmed Earth's magnetosphere, allowing the particles to reach the lower layers of the atmosphere. This "magnetic storm" started fires, disrupted telegraphic communications, and produced light displays that were observed even in tropical areas such as Cuba and Hawaii.

through a magnetic field is deflected, because the magnetic force alters its path by pushing the charge sideways. If the magnetic force is strong enough, the electric charge will travel in a circle!

Magnetic forces acting on electric charges are important for Earth because of the presence of charged particles in space and in the upper atmosphere. Although the space between the planets in the solar system is mostly empty, it contains a small number of particles, including some having an electric charge. Some of these particles come from the Sun, which emits a stream of hot gas traveling at 1,000,000 miles (1,600,000 km) per hour called the solar wind. (High temperatures and other, less understood effects boost the speed of these particles, allowing them to escape the Sun's gravity.) Much of the solar wind consists of electrons and protons. Another source of charges occurs when high-energy radiation from the Sun strikes atoms in Earth's upper atmosphere, stripping electrons and producing *ions*—electrically charged particles.

When these charged particles encounter Earth's magnetosphere, their paths are altered. Earth's magnetosphere is also affected—recall that charges in motion generate magnetic fields. As a result of the interactions between Earth's magnetic field and these charges, the planet's magnetosphere is not spherical in shape, as the name suggests, but is "swept" slightly away from the Sun by the solar wind.

The altered shape of the magnetosphere is by no means the most crucial aspect of this interaction. If the particles of the solar wind were not deflected by Earth's magnetosphere, they might strike deep within the atmosphere or on the planet's surface. A high-speed particle can cause a lot of damage to a molecule or even break it apart in a collision. Large biological molecules, such as the molecules composing much of the human body, are especially vulnerable. Constant bombardment by the solar wind might seriously impair living organisms. The magnetosphere helps protect against these adverse effects, but Earth's dense atmosphere is also critical because it blocks many of these particles from reaching the surface of the planet.

Other consequences of the interaction between Earth's magnetic field and charged particles include colorful and eerie displays of light. As described in the sidebar on page 44, these lights appear in the sky in extremely northern and southern latitudes, especially around the poles. Accelerations in the speed of electric charges, created by magnetic interactions, and collisions with gaseous molecules in the atmosphere are responsible for these beautiful displays.

Mapping the magnetosphere requires venturing into space. Satellites provide one of the best means to study the extension of Earth's magnetic field in space, giving geologists a bird's-eye view. To make an

Magnetometer used to measure magnetic fields in three dimensions—note the instrument's three axes *(NASA/CETP)*

even more precise set of measurements, the European Space Agency (ESA) plans to launch a group of three satellites in 2010. This mission, called Swarm, aims to place each of the three satellites in a different polar orbit—an orbit in which the satellite's path carries it over the poles—at an altitude in the range of 248–341 miles (400–550 km). Instruments such as *magnetometers* will measure the strength and direction of the magnetic field, along with additional equipment such as accelerometers to study the interactions with electrically charged particles.

DYNAMO THEORY OF EARTH'S MAGNETIC FIELD

The internal heat of Earth and the properties and variability of the geomagnetic field strongly suggest that the notion of Earth as a bar magnet is useful but not precisely correct. Earth's magnetic field does not come from an oriented bar magnet buried underneath the surface.

Yet Earth's iron-rich core offers other possible explanations. Recall that the core has two sections: an inner core made of mostly solid iron and a liquid outer portion of iron along with nickel and a few lighter elements. Although the core's high temperatures preclude ferromagnetic properties, the heat creates currents in the liquid outer core. These currents are often called convection currents because they involve a heat transfer mechanism known as convection, in which the movement of fluid carries heat from one place to another. A breeze, for example, is a convection current that helps people to cool down on a hot day. Earth's convection currents are part of the means by which the planet cools off. Convection currents in the mantle also help explain plate tectonics, and the currents in the outer core are even stronger because it is much more fluid than the mantle.

Most geologists believe the origin of Earth's magnetic field is due to a dynamo effect. Such ideas have been around since the early 20th century, when the Irish scientist Sir Joseph Larmor (1857–1942) proposed a dynamo theory to explain certain magnetic properties of the Sun. A dynamo is a machine that generates electricity by spinning an electrical conductor in a strong magnetic field. (The motion of the conductor causes the magnetic field in the conductor to vary, which induces an electric current.) Iron is an effective electrical conductor, and its rapid motion due to convection currents, along with the rotation of the planet, may set up a kind of dynamo within Earth. This process induces an electric current, which as Oersted discovered, creates a magnetic field. If the dynamo theory is correct, Earth's internal heat is critical for its magnetic field, because the heat creates convection currents to drive the necessary motion of the conductor.

But a question arises concerning this theory. The dynamo generates an electric current that in turn produces a magnetic field, but a magnetic field is necessary in the first place in order for the dynamo to generate an electric current! If convection currents in the liquid core act as a dynamo, geologists still must account for the magnetic field required for the dynamo to operate.

This requirement of a magnetic field for the dynamo need not pose a difficult obstacle for the theory, however. Weak magnetic fields occur in many materials. These weak fields may come from electric currents arising from some mechanism in the iron core, possibly due to its structure or extreme pressure and temperature. Although these fields are not strong enough to explain Earth's magnetic field, the dynamo in the liquid outer core amplifies these fields.

As is true for almost all theories at the frontiers of science, the dynamo theory is not simple to prove. Seismic waves give geologists clues about the inner structure of the planet, but no one can open up Earth's interior and peer inside. Earth's magnetic field probably comes from a dynamo in Earth—sometimes referred to as the geodynamo—but nobody can be certain.

Although the assurance of certainty eludes scientists, evidence supporting the theory mounts. One way to test a theory is to use computers to simulate the situation. Chapter 1 discussed some of the ways in which fast computers known as supercomputers are important in geology, and the dynamo theory is another opportunity for supercomputers to become involved in Earth science. In the 1990s scientists such as Gary Glatzmaier of Los Alamos National Laboratory in New Mexico, Paul Roberts at the University of California, Los Angeles, and Jeremy Bloxham and Weijia Kuang at Harvard University developed computer models showing how the geodynamo might work.

The problem is that these models are quite different. For example, one of the models suggests that the magnetic field arises from the lower depths of the outer core, while another posits that the field is produced by mechanisms existing in the upper regions of the outer core. Both models account for the properties of Earth's magnetic field, as measured from space and the surface of the planet.

As the speed and sophistication of computers advance, simulations include more details and are becoming increasingly realistic. Futoshi Takahashi at the Japan Aerospace Exploration Agency and Masaki Matsushima and Yoshimori Honkura of the Tokyo Institute of Technology recently formulated a detailed model of the dynamo process. The researchers used one of the fastest supercomputers in the world—the Earth Simulator, a computer at the Japan Agency for Marine-Earth Science and Technology designed to study large systems. The researchers published their report, "Simulations of a Quasi-Taylor State Geomagnetic Field Including Polarity Reversals on the Earth Simulator," in a 2005 issue of *Science.*

Earlier models did not faithfully replicate the core's dynamics and properties such as those involving viscosity (the ease of flow). In the report, Takahashi and his colleagues noted, "It is still unclear whether these models accurately reproduce the dynamics of Earth's outer core, because the simulations were performed at a dynamic regime much different from that of the core." To construct the newer model, the researchers matched the measured and theoretical properties of Earth's magnetic field even more closely than before. The predictions such simulations make about the future behavior of the magnetic field, including fluctuations and reversals of the poles, are of great importance, as will be discussed more in the final three sections of this chapter.

Other clues about the origin and nature of magnetic fields need not be confined to sources close to home. The solar system contains many other large objects, some of which also exhibit magnetic behavior.

MAGNETIC FIELDS OF OTHER BODIES IN THE SOLAR SYSTEM

Venus is the second planet from the Sun. Its orbit is closer to the Sun than the orbit of Earth, the third planet, by about 25 million miles (40 million km). Venus resembles Earth in many ways, including size—the radius of Venus is only 5 percent smaller—and density, which also differs by only a few percent. But the temperature of the surface of Venus is much warmer than Earth, due to the proximity of Venus to the Sun as well as its thick atmosphere of carbon dioxide, which traps heat. The length of Venus's day, as determined by the time it takes for a planet to complete one revolution on its axis, is also considerably different. Earth's day lasts 24 hours, but a day on Venus lasts 243 times as long as that of Earth—Venus rotates so slowly that it takes about 5,830 hours for it to finish one spin!

Geologists began to suspect Earth has an iron core because its density is much higher than would be expected if the planet were composed entirely of rock. The study of seismic waves supported this notion. Although seismic data from Venus is not available, the planet's density suggests it has an iron core similar to Earth. Venus's interior is also likely to be as hot as Earth; the planets are similar in size and age, so both have retained some of the heat of their fiery birth. Venus may be even hotter, due to its high surface temperature and proximity to the Sun.

The similarities between the two planets suggest that the dynamo theory might well apply to Venus as it does to Earth. But space probes such as

Mariner 2, which flew past Venus in 1962, have failed to detect a magnetic field coming from the planet—Venus is not very magnetic, if at all.

A lack of magnetism from Venus would seem to contradict the application of the dynamo theory to planetary magnetic fields, yet there are important differences between Venus and Earth that might explain the discrepancy. Venus seems to lack dynamics such as plate tectonics—the motion of rigid plates of crust driven by mantle convection—and the planet rotates extremely slowly. Convection currents may not be well organized within the planet, or the snail-like pace of its rotation may be too slow to stir them up sufficiently.

Mars is another planet that probably has a metallic core, but like Venus, it lacks any significant magnetic field. Fourth from the Sun, Mars's orbit is about 45 million miles (72 million km) farther than Earth's orbit. The rotation rate of Mars is similar to Earth—a Martian day is about 24.5 hours. Scientists are unsure whether the core of Mars is solid or liquid, but in any case the planet may be too cool to support internal convection currents. Yet the space probe *Mars Global Surveyor,* which arrived at the planet in 1997 and mapped the surface from a low-altitude orbit, signaled the presence of ancient magnetized rocks. This suggests the planet possessed a substantial magnetic field early in its history, which would support the dynamo theory if the state of Mars in its early stages was conducive to such a process—and the planet certainly would have been hotter.

Mercury turns out to be a bit more perplexing. Although Mercury is the closest planet to the Sun, it is small—the radius is only 38 percent that of Earth—and such a small planet might have a completely solid core. Mercury's rotational rate—58.8 times slower than Earth—also appears far too sluggish for a dynamo to exist. Yet when the space probe *Mariner 10* encountered Mercury in 1974 and 1975, it found a weak but measurable magnetic field!

Scientists are uncertain what to make of Mercury's magnetic properties. In its simplest conception, the dynamo theory fails to explain this planet's magnetism. Yet perhaps the closest planet to the Sun is more mysterious than it appears. The core may be molten, or partially molten, and the proximity to the Sun's enormous gravity may be creating powerful internal forces to circulate the fluid. If so, the dynamo theory may explain Mercury's magnetic field as successfully as it seems to account for Earth's.

To answer the questions about planetary cores and magnetism, probes with seismometers must land on the surface. These probes would not only address issues concerning magnetism, but also reveal a great deal of information of general interest to geologists. Writing about Mars in a 2005 issue of *Science,* the researchers Yingwei Fei and Constance Bertka of the Carnegie Institution of Washington commented, "Space missions with multiple landers equipped with seismometers are required to precisely determine the size of the core. Such missions would also provide fundamental information on the structure and density profile of the martian interior, which is critical for understanding both the formation and evolution of Mars. This understanding is essential for providing a general context to explore the formation and evolution of terrestrial planets, including our own." The same may be said for missions to other planets. But space missions are expensive and must compete for funds with other important projects, so these questions have not yet been addressed.

Other bodies of the solar system, such as Jupiter and the Sun, have strong magnetic fields. But Jupiter and the Sun are much larger than the inner planets of the solar system and have vastly different internal structures. Jupiter is mostly hydrogen, possibly with a small rocky or metallic core. The Sun is a sphere of hot gas with a volume that would contain about 1,300,000 planets the size of Earth. Some of these gases consist of ions, and their movements generate the Sun's enormous and complex magnetic field. Jupiter's magnetic field probably has a similar origin.

VARIATIONS IN EARTH'S MAGNETIC FIELD

The Sun's gaseous structure permits a lot of fluctuations. For instance, disturbances such as the one in 1859, described in the sidebar on page 45, transiently altered its properties, and the magnetic field of the Sun is also extremely variable. Earth is rocky and much less fluid, but the magnetic field of Earth is also subject to considerable variability.

The experiments by Tarduno and his colleagues on 3.2-billion-year-old rocks indicate that Earth's magnetic field was about 50 percent as strong then as it is now. But the strength of the magnetic field has changed in recent times as well.

Measurements of magnetic fields rely on the interactions of these fields with other magnets, ferromagnetic materials, or electric charges. Iron filings orient themselves along the lines of force that a magnetic field exerts, for example, as does a compass needle exposed to Earth's magnetic field. A magnetometer provides more information by measuring the strength and direction of a magnetic field. The German mathematician and scientist Carl Friedrich Gauss (1777–1855) published the earliest description of a magnetometer in 1833; the instrument was simple, consisting of a bar magnet suspended by a gold fiber, and indicated magnetic field strength by the period of its oscillation. (Magnetic fields affect the time, or period, of a magnet's back and forth motion.) One of the standard units of magnetic field force or intensity is named after Gauss to honor his achievements.

Instruments such as SQUIDs are much more sensitive than Gauss's simple magnetometer, and many other types of magnetometer have been developed. But any use of magnetometers to measure Earth's magnetic field must take into account possible contributions by "stray" magnetic fields. Sources for these fields include currents flowing in electrical equipment and especially the large amount of current flowing in the overhead or underground wires of the power companies. Even iron that is not magnetized may have small, disorganized magnetic fields due to partial alignment of domains. (This feature of iron troubled a lot of navigators and sailors in the 19th century as shipbuilders switched from wood to steel, which is mostly iron—stray magnetic fields in the ship's steel disrupt the compass unless the instrument is adequately shielded.)

Despite the difficulties, records of careful measurements date back to the 1840s, when sailors and scientists began charting Earth's magnetic field strength. These measurements continue today at stations such as Hartland Magnetic Observatory, operated by the British Geological Survey (BGS) and located in a large meadow in southwestern England near the village of Hartland in the county of Devon. By comparing past records with present measurements, geologists have found a decline of roughly 10 percent in Earth's magnetic field strength in the last century and a half.

Analysis of rocks lets scientists explore the history of Earth's magnetic field even further back in time. *Paleomagnetism* refers to the preservation or record of Earth's magnetic field in the structure of rocks and other geological formations. Studies such as that of Tarduno and his colleagues reach back billions of years.

One of the most important branches of paleomagnetism involves the study of iron-bearing igneous rocks, which are formed by the cooling and solidification of molten rock. Lava eruptions, for example, cool and harden over time. The presence, strength, and direction of Earth's magnetic field at the time these rocks solidify has a powerful influence on the orientation of the particles of iron; while the rock is molten, the iron and iron minerals orient to the field, and become fixed in place as the rock crystallizes. Contained within these rocks is a sketchy but informative history of Earth's magnetic field.

The most startling finding of paleomagnetism comes from studies of the seafloor. In the 1940s during World War II, military engineers from the United States and other countries developed magnetometers to scan for submarines. These magnetometers, towed by ships, detected magnetic fields associated with the steel hulls of submarines. After the war, geologists modified this equipment to study the ocean floor and were surprised to discover regular variations in magnetic fields as they sailed back and forth across the ocean, particularly the middle of the Atlantic Ocean. These field fluctuations were natural phenomena, although anomalous—unexpected and unexplained.

Other geological observers were mapping the seafloor with sonar—sound navigation and ranging—which uses reflected sound waves to locate objects underwater. Geologists found vast ridges cutting across the oceans that contained numerous underwater volcanoes. As described in the following sidebar, mid-ocean ridges form as two tectonic plates gradually slide apart, creating a rift or gap that quickly fills with an upwelling of magma.

The molten rock solidifies to become a new portion of Earth's crust. Magnetic orientation of this new crust reflects Earth's magnetic field at the time of solidification. As the plates slowly move apart, the new crust moves with it, opening the gap again for molten rock. The process repeats, with the newest crust forming in the center, and the slightly older crust shifting to each side. The seafloor along these ridges acts as a magnetic tape recorder—a record of the history of the magnetic field. This record indicates that not only has the strength of Earth's magnetic field varied, the direction of the field has varied as well.

Mid-Ocean Ridges

In 1912 the German scientist Alfred Wegener (1880–1930) made his astonishing proposal that continents drift. Skeptical geologists could not imagine how something as large and heavy as a continent could "set sail," but in the 1950s researchers discovered long, ocean-girdling ridges. A ridge is a range of hills or hilly terrain, but the ocean ridges also contained a crack—a deep valley or rift. In the 1960s the researchers Harry Hess (1906–69) at Princeton University and Robert Dietz (1914–95) at the Scripps Institution of Oceanography suggested that these rifts form when rigid crustal plates separate. Hot magma from below rises up to create a new section of the seafloor.

Seafloor spreading was a critical factor in the establishment of plate tectonic theory. Plate boundaries form where plates meet. In some cases one plate slides past or dips beneath another plate, but in other cases two plates separate, or diverge. Divergence is what occurs at mid-ocean ridges, which are plate boundaries in which two plates slowly move apart.

For example, the Mid-Atlantic Ridge runs along the middle of the Atlantic Ocean, created by the separation of the North American and South American Plates from the African and Eurasian Plates. These plates separate by about 0.4 inches (1 cm) every year. (Part of the valley cuts through Iceland and is aboveground!) This region is a geologically active site; nearly all earthquakes in the Atlantic Ocean occur at or near the Mid-Atlantic Ridge. Plenty of young crust is available for inspection as well.

As geologists obtained samples from the seafloor from the mid-ocean ridges, they found the ages of the rocks corresponded to the movement of the plates—the newest and youngest rocks are at the rift center, where the plates are presently diverging, with a progressively increasing age farther away from the rift, at sites that formed longer ago.

POLE REVERSALS

The anomalous magnetic patterns that geologists observed from their sea magnetometers came from bands of rocks with opposing magnetic orientations. During some periods of time in Earth's history, the mag-

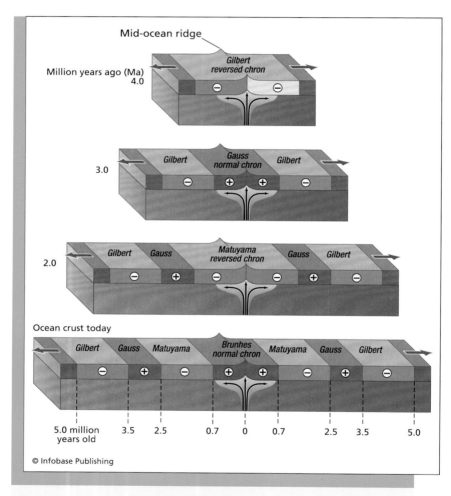

© Infobase Publishing

The spreading seafloor contains a record of pole reversals—as the seafloor spreads, rocks having one or the other magnetic orientation form bands or chrons. The most recent chrons, from oldest to newest, are Gilbert, Gauss, Matuyama, and Brunhes (all named for scientists who made important contributions to the study of geomagnetism).

netic poles have been as they are now. In other periods of time, the poles have been reversed—the south magnetic pole was in the north! The effect is the same as that achieved by rotating a bar magnet by 180 degrees; if the north magnetic pole was oriented upward in the original position, then after the rotation it would point downward.

Geologists have also taken a close look at other sites, such as periodic lava flows of land volcanoes. Decades before scientists discovered the mid-ocean system, some researchers, such as the French scientist Bernard Brunhes (1867–1910), had found evidence of magnetic reversal in rocks (though the idea did not gain wide acceptance at the time, since other explanations were possible). Later, scientists determined that the findings on land corresponded with those of the mid-ocean ridge. About one-half of rocks that have been examined show a different magnetic orientation, which is the reverse of the field's present orientation. At various points in time, Earth's magnetic poles have switched places. The time between these reversals, during which the field has one or the other orientation, is known as a *magnetic chron*. Chrons that have been documented in the spreading seafloor can be seen in the figure on page 56.

Magnetic pole reversals are also evident in the Sun and its enormous magnetic field. The Sun's magnetic field reverses every 11 years. This regular cycle is associated with other phenomena such as sunspots—areas of lower temperature and greater magnetic activity—but astronomers do not fully understand these processes.

On Earth magnetic reversals are much less predictable. The "tape recorder" written on rocks stretches back millions of years, and major chrons last about 500,000 years on average. But the reversal rate has varied widely. In the Cretaceous period, 145 to 65 million years ago—when dinosaurs lived and roamed the planet—reversals were less frequent than they are now, sometimes millions of years passed between reversals. For the last 10 million years, reversals have come more often, averaging about five reversals per million years. The last reversal occurred about 780,000 years ago.

How much time does Earth's magnetic field take to reverse poles? The magnetic record is not precise down to small periods of time such as a month or a year, since other processes affect rocks during their formation as well as afterward, and geologists cannot make exact statements. There appears to be variability in the amount of time each of these transitions last, ranging from 1,000 years to almost 30,000 years. The field strength recedes to an extremely low quantity, perhaps about 10 percent of its normal value, and then gradually regains its strength with the poles reversed.

Bradford M. Clement, a researcher at Florida International University, recently analyzed records from sediment cores. Scientists obtain these cores by drilling in sediments such as those on the ocean floor. The drill is hollow, and as it cuts through mud and sedimentary rock it generates a cylindrical sample, or core, which can be retrieved and studied. The sediment layers are in chronological order, and the deepest layers, being the first to settle, are the oldest. Clement studied the four latest magnetic field reversals as recorded in sediment cores from different parts of the world and found the transition lasted an average of 7,000 years.

But the variations in transition time had an interesting feature. In cores at low latitudes, near the equator, the transition occurred more quickly than at higher latitudes. For example, the last reversal was over in about 2,000 years at the equator but lasted 8,000 years in more northern latitudes. Clement published his report, "Dependence of the Duration of Geomagnetic Polarity Reversals on Site Latitude," in a 2004 issue of *Nature*.

The variation in transition times at different latitudes on the globe may be due to the geometry of the geodynamo and the flow of currents in the inner core. But the important question of why these reversals occur at all remains unanswered.

As is true for many other topics in geology, especially interior processes, models and computer simulations are important tools to investigate the cause or causes of magnetic pole reversals. As scientists ponder these reversals, two competing ideas have emerged. One idea is that the reversals are due to some internal event within the convection currents and magnetic activity of the inner core. The other idea is that some external event, outside of Earth's interior, triggers a pole reversal by interfering or influencing the outer core.

Richard A. Muller, a physicist at the University of California, Berkeley, has suggested a possible role for comet or asteroid impacts as an external trigger for pole reversals. When a comet or asteroid strikes Earth's surface at an oblique angle—sideways instead of straight on—the impact exerts a twisting or shearing force that can be felt all the way to the core-mantle boundary. Such an event might cause an "avalanche" of rocky material and lighter elements that compose the boundary between the rocky lower mantle and the liquid outer core. If these disturbances are severe enough, convection currents within the outer core may become disrupted, temporarily halting the dynamo and perhaps even reversing the polarity of the magnetic field it creates. Muller published some of his ideas in a paper, "Avalanches at the Core-Mantle Boundary," in a 2002 issue of *Geophysical Research Letters.*

But is an external trigger necessary? The computer simulation of Takahashi and his colleagues, published in *Science* in 2005, suggests pole reversals are a natural phenomenon having no need for external causes. Takahashi and his colleagues observed magnetic reversals in the simulations when regions of magnetic activity drifted toward the poles and disappeared, replaced by activities in the middle that have a different field orientation. These findings suggest that the duration of the reversal depends on latitude, an observation earlier made by Bradford M. Clement, as described above.

Computer simulations that show pole reversals despite the absence of external triggers suggest reversal is a spontaneous process—a naturally occurring feature of the geodynamo. But this question is not yet decided. Even with the fastest supercomputers, such as Earth Simulator Center used in the simulations of Takahashi and his colleagues, the programs are only a rough approximation of the complexity of Earth's core and its surroundings. Although progress has been made, further progress requires even faster computers and more detailed simulations. And the details for these simulations must come from additional experimental or observational data, such as that of Clement, in order for geologists to construct a more complete model of Earth's hidden and complicated magnetic field generator.

CONCLUSION

Compass needles point toward the magnetic pole and have guided intrepid sailors for hundreds of years. Yet even as Gilbert was formulating his ideas of Earth as a magnet, observers began noticing variability in the forces acting on compasses. Earth's magnetic field varies, with continually shifting magnetic poles and the occasional pole reversal.

The records described above show a reduction in the strength of Earth's magnetic field compared to 150 years ago. Such a decline in magnetic field strength is probably one of the hallmarks of a pole reversal, which leads many people to wonder if Earth may be in the midst of another change. Another reversal could be overdue—reversals have occurred in the past at an average of a few hundred thousand years, and the last one clocked in at 780,000 years ago.

In the absence of a complete understanding of how pole reversals occur, scientists cannot be certain if the observed magnetic field decrease is the beginning of a transition or just a temporary variation unrelated to any reversal process. But in a FOX News report by Clara Moskowitz on September 26, 2008, Brad Singer, a researcher at the University of Wisconsin, said that "we might experience a field reversal in the next two millennia if it continues to weaken at the current rate."

The issue is an important one. Should a reversal begin to occur in the near future, disruptions in navigation and communication could ensue. The health and welfare of animals as well as human beings could also be affected, given the roles that the magnetosphere plays in shielding the planet.

But according to the geological record, as written in volcanic rocks and mid-ocean ridges, a reversal takes thousands of years. Scientists are unsure of the nature and sequence of events accompanying these transitions—this knowledge is one of the primary goals of computer models and simulations—but the planet and the organisms living on its surface and beneath its waters have certainly survived such transitions many times in the past. "The magnetic field has reversed itself hundreds, if not thousands, of times," noted Singer. And there is no confirmed fossil evidence linking pole reversals with mass extinctions.

Although declines in magnetic field strength accompany pole reversals, the field does not entirely disappear. Life on Earth also has a thick atmosphere for protection. If a reversal does actually take place in the near future, some disruptions may be expected, but will probably not be severe. There is comfort in precedents—the ancestors of modern humans survived pole reversals. *Homo erectus,* who lived from about 2,000,000 years ago to 500,000 years ago, endured the last reversal.

Geologists may soon discover models or make observations that will determine if Earth is approaching another magnetic reversal. These findings would also reveal a great deal of other information about Earth and its surroundings. What goes on deep inside the planet affects the surface as well as space, which makes the topic a vital, and fascinating, frontier of science.

CHRONOLOGY

ca. 778,000
B.C.E.
The poles of Earth's magnetic field reverse, settling in their present configuration.

ca. 1000 C.E.   Chinese navigators develop the compass.

1600
The British scientist William Gilbert (1544–1603) publishes *De Magnete* (On the magnet), a book in which he proposes that Earth behaves as a magnet.

1820
The Danish physicist Hans Christian Oersted (1777–1851) discovers that an electric current generates a magnetic field.

1831
The British scientist Michael Faraday (1791–1867) observes that a changing magnetic field induces an electric current in a conductor.

The British explorer Sir James Clark Ross (1800–62) reaches the north magnetic pole.

1833
The German scientist Carl Friedrich Gauss (1777–1855) publishes the earliest description of a magnetometer.

1904
The Norwegian explorer Roald Amundsen (1872–1928) becomes the second person to find the northern magnetic pole, but he discovers it had moved about 30 miles (50 km) north of Ross's location.

1906
The British seismologist Richard D. Oldham (1858–1936) analyzes seismic waves to show that part of Earth's core—the outer core—is liquid.

The French scientist Bernard Brunhes (1867–1910) discovers rocks that suggest the poles of Earth's magnetic field have been reversed in the past.

1919	The Irish physicist Sir Joseph Larmor (1857–1942) proposes a dynamo theory to explain some of the Sun's magnetic properties.
1960s	Harry Hess (1906–69) and Robert Dietz (1914–95) propose the crust is made of rigid plates that are separating to form mid-ocean ridges. Rocks surrounding these ridges have stored a magnetic record of the history of Earth's magnetic field.
1974	Space probe *Mariner 10* flies past Mercury and detects a weak but significant magnetic field. This field is hard to explain in terms of the dynamo theory.
1990s	Gary Glatzmaier, Paul Roberts, Jeremy Bloxham, and Weijia Kuang develop computer models of a dynamo-like process in the inner core to explain Earth's magnetic field properties.
2000s	Models and computer simulations suggest that Earth's magnetic pole reversals are due to internal processes.
2008	Signs of weakening in the Earth's magnetic field may indicate another pole reversal could occur soon.

FURTHER RESOURCES

Print and Internet

Aczel, Amir D. *The Riddle of the Compass: The Invention That Changed the World.* New York: Harcourt, 2002. For hundreds of years people have used the compass to navigate the treacherous seas. The story unfolds in this concise book, as Europeans emerge from their reclusion of the Middle Ages and set out to explore the world.

Campbell, Wallace H. *Earth Magnetism: A Guided Tour through Magnetic Fields.* New York: Academic Press, 2001. This book takes the reader on a complete and fascinating tour of Earth's magnetic field and its properties.

Clement, Bradford M. "Dependence of the Duration of Geomagnetic Polarity Reversals on Site Latitude." *Nature* 428 (April 8, 2004): 637–640. Clement reports on his studies of the four latest magnetic field reversals, as recorded in sediment cores.

European Space Agency. "Swarm." Available online. URL: http://www.esa.int/esaLP/LPswarm.html. Accessed May 4, 2009. Swarm, an ESA project, will launch a set of satellites to survey Earth's magnetic field with unrivaled precision. This Web resource describes the mission, objectives, and participating satellites.

Fei, Yingwei, and Constance Bertka. "The Interior of Mars." *Science* 308 (May 20, 2005): 1,120–1,121. The authors call for a planetary mission to study the interior of Mars.

Geological Survey of Canada. "Geomagnetism." Available online. URL: http://gsc.nrcan.gc.ca/geomag/index_e.php. Accessed May 4, 2009. The Geological Survey of Canada is heavily involved in tracking the north magnetic pole and other projects concerning Earth's magnetic field. The Web resource contains links to information on geomagnetic observatories, the north magnetic pole, geomagnetic forecasts, and similar topics.

Gilbert, William. *On the Magnet.* Translated by Silvanus Thompson. Available online. URL: http://rack1.ul.cs.cmu.edu/is/gilbert/. Accessed May 4, 2009. This resource contains a translation of Gilbert's classic work.

Glatzmaier, Gary A. "The Geodynamo." Available online. URL: http://es.ucsc.edu/~glatz/geodynamo.html. Accessed May 4, 2009. Glatzmaier, a researcher interested in Earth's magnetic field, provides a well-written description of simulations aimed at exploring and understanding the field's origin.

Livingston, James D. *Driving Force: The Natural Magic of Magnets.* Cambridge, Mass.: Harvard University Press, 1996. Magnets and magnetism have interested people ever since the ancient Greeks. Livingston relates the history of research on magnets and magnetic fields and explains how magnetism works.

Moskowitz, Clara. "Earth's Magnetic Field Expected to Flip Soon." Fox News, September 26, 2008. Available online. URL: http://www.

foxnews.com/story/0,2933,428849,00.html. Accessed May 4, 2009. Some researchers believe a magnetic pole reversal may occur in the near future.

Muller, Richard A. "Avalanches at the Core-Mantle Boundary." *Geophysical Research Letters* 29 (2002): 1,935–1,939. The author suggests a possible role for comet or asteroid impacts as an external trigger for pole reversals.

National Aeronautics and Space Administration. "What Is the Magnetosphere?" Available online. URL: http://science.nasa.gov/ssl/pad/sppb/edu/magnetosphere/. Accessed May 4, 2009. This accessible series of Web pages explains the magnetosphere and includes numerous illustrations.

National Geophysical Data Center. "Geomagnetic Field Frequently Asked Questions." Available online. URL: http://www.ngdc.noaa.gov/geomag/faqgeom.shtml. Accessed May 4, 2009. The National Geophysical Data Center manages data accumulated by various geological, oceanic, and atmospheric observatories. The frequently asked questions (FAQ) listed on the Web page cover the basic principles of Earth's magnetic field, along with such topics as compasses and magnetic pole reversals.

Public Broadcasting Service. "Magnetic Storm." Available online. URL: http://www.pbs.org/wgbh/nova/magnetic/. Accessed May 4, 2009. *Nova,* a science documentary series broadcast on PBS, aired a program in 2003 on the possibility that Earth's magnetic poles will reverse in the near future. This Web resource is an informative companion to this program, and includes a simulation of a pole reversal.

ScienceDaily. "3.2 Billion-Year-Old Surprise: Earth Had Strong Magnetic Field." News release, April 5, 2007. Available online. URL: http://www.sciencedaily.com/releases/2007/04/070404162406.htm. Accessed May 4, 2009. This news release describes the article that John A. Tarduno and his colleagues published in *Nature* 446 (April 5, 2007): 657–660.

Stern, David P. "The Great Magnet, the Earth." Available online. URL: http://www.phy6.org/earthmag/demagint.htm. Accessed May 4, 2009. William Gilbert introduced the idea that Earth is a giant magnet in his 1600 book, *De Magnete* (On the magnet). This Web resource includes links to essays discussing this book, Gilbert's ideas, and subsequent developments in the study of magnetism.

Takahashi, Futoshi, Masaki Matsushima, and Yoshimori Honkura. "Simulations of a Quasi-Taylor State Geomagnetic Field Including Polarity Reversals on the Earth Simulator." *Science* 309 (July 15, 2005): 459–461. The researchers report a model that matches the measured and theoretical properties of Earth's magnetic field more closely than earlier models.

Tarduno, John A., Rory D. Cottrell, Michael K. Watkeys, and Dorothy Bauch. "Geomagnetic Field Strength 3.2 Billion Years Ago Recorded by Single Silicate Crystals." *Nature* 446 (April 5, 2007): 657–660. These results indicate Earth's magnetic field was about half as strong as it is today.

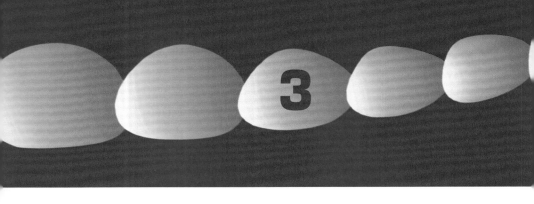

VOLCANOES AND HOT SPOTS

Ancient Greek poets wrote about a god of fire known as Hephaestus, who forged metallic objects in the furnace beneath the Italian volcano Mount Etna, located on the east coast of Sicily. The Romans called this god Vulcan, or, in Latin, *Volcanus*, from which the term *volcano* derives.

Myths of Hephaestus or Vulcan did not advance volcanology—the study of volcanoes—but some ancient Greek philosophers tried to formulate a rational explanation for these powerful geological phenomena. Empedocles (ca. 490–30 B.C.E.), a philosopher and citizen of a Greek colony in Sicily, proposed a theory in which the world consists of four elements—earth, water, air, and fire. Volcanoes were important examples of fire. (According to a colorful but doubtful legend, Empedocles died when he fell or threw himself into a fiery pit on Mount Etna.) Although the four-element philosophy gained some adherents in the ancient world and during the Middle Ages, it made little contribution to a scientific understanding of volcanoes.

One of the reasons why geologists study these violent phenomena is because volcanic eruptions are so dangerous to people living nearby. But volcanoes also provide clues about the structure and properties of the planet. Much volcanic activity is associated with the boundaries of tectonic plates, but a few volcanoes have another cause—the activity of these volcanoes seems to be due to small, hot regions, called hot spots, in the middle of a plate. This chapter describes the nature of volcanoes and focuses on the efforts of researchers to understand hot spots.

INTRODUCTION

In 79 C.E. Pompeii was a bustling city located near the Bay of Naples in a region of western Italy known as Campania. Founded by an Italian people called the Oscans centuries earlier, the Romans had conquered and annexed the city by about 290 B.C.E. The soil in Campania was rich in nutrients and extremely fertile. Many wealthy Romans lived in or around the Bay of Naples or spent long vacations there in luxurious villas, and the population of Pompeii was about 10,000 to 20,000 people. But on August 24, 79 C.E., a previously inactive volcano on nearby Mount Vesuvius erupted, covering the region in lava and ashes—and burying Pompeii.

About 2,000 people in Pompeii died, blanketed in more than 30 feet (9.1 m) of ash, rock, and debris. Many people outside of the city also perished, including Gaius Plinius Secundus (ca. 23–79 C.E.)—better known as Pliny the Elder—the commander of a nearby Roman naval base. In addition to his military duties, Pliny the Elder was a prolific author who was fascinated with history and nature; when Vesuvius began erupting, he sailed across the bay to investigate. But the scientific expedition quickly evolved into a rescue operation as the extent of the disaster became clear, and Pliny the Elder died while helping the evacuation. Pliny's nephew, Pliny the Younger (ca. 62–113 C.E.), had stayed behind at the naval base and later wrote about the tragedy. This writing has survived to the present day, collected in *The Letters of Pliny the Younger.*

Excavators digging in the area rediscovered the buried city in the middle of the 18th century. The ash and debris preserved many of the structures from the ravages of time, and the site is one of the most popular tourist attractions in Italy, offering an unsurpassed glimpse of what life was like in an ancient Roman city. The tragedy at Pompeii also serves as a warning, as do other eruptions that have caused much destruction, such as the 1980 eruption of Mount St. Helens in Washington, which killed 57 people, and the 1985 eruption of Nevado del Ruiz in Colombia, which claimed about 25,000 lives and covered hundreds of square miles of land.

Pliny the Younger's description of the eruption of Mount Vesuvius was one of the earliest recorded observations of volcanic activity. Vesuvius is also the site of the earliest modern volcano observatory, established around 1847. This was the only volcano observatory in the world for many years, until a catastrophe occurred on Martinique, an island in the Caribbean Sea. On May 8, 1902, Mount Pelée erupted, destroying the island's city of Saint-Pierre and killing 30,000 people. The disaster prompted the construction of an observatory to monitor this volcano. Although the volcano soon became inactive and the observatory was abandoned, subsequent volcanic activity in the late 1920s drew additional interest, including by the American engineer, inventor, and volcanologist Frank A. Perret (1867–1943), who helped establish a permanent observatory on Martinique. Throughout his long career, Perret was involved in a great deal of research at many different volcanoes.

An article on Perret, written by Mildred Giblin and published in 1950 in the *Bulletin of Volcanology,* describes how Perret got involved in the study of volcanoes: "It was more or less by accident that Mr. Perret entered this field of research, his early training and interest having been in the development of electrical instruments for industrial application. Well launched in commercial enterprises, Perret, as the result of a severe illness, was compelled to take a complete rest. The necessity of this enforced idleness served, however, to bring him into a new world of men and thought which soon disclosed to him his life work and brought him international scientific renown." Perret, who had studied physics at the Brooklyn Polytechnic Institute (now called Polytechnic University) in New York, vacationed in Italy in the early 20th century, and a visit to Vesuvius sparked his interest in volcanoes. He spent the next 30 years as a volcanologist, tirelessly making observations and taking photographs.

Observations of Perret and many other researchers led to the scientific description and analysis of volcanoes. A volcano is an opening in the surface of Earth from which molten rock, gases, and other material may erupt from time to time. Molten rock is called magma when it is beneath the surface and lava when it erupts from a volcano. The lava cools as it solidifies and gradually forms a hill or mountain surrounding the volcanic opening.

Some volcanoes are more active than others, which means they erupt more frequently. In the United States some of the most active volcanoes are Kilauea in Hawaii and Mount St. Helens. Other volcanoes have failed to erupt for long periods of time and are called *dormant*. The total number of active volcanoes in the world ranges from about 500 to 1,500, depending on who is doing the counting and how active a volcano has to be in order to make the list. For example, Yellowstone National Park in Wyoming is the site of a large volcano that last erupted

A 1969 eruption of Kilauea volcano in Hawaii *(D. A. Swanson/USGS)*

about 640,000 years ago. Although a long time has passed since the last major eruption, some geological activity, as described below, is evident from the recent past and even today, leading some people to consider the volcano an active one.

Volcanoes come in different shapes and sizes. Some volcanoes have a central opening or vent that is connected to a magma chamber—an underground "lake" of molten rock—by way of a channel. Magma and gases ascend through this channel and through the vent during an eruption.

The shape of a central opening depends on the nature of the ejected material. If the eruptions mostly consist of a type of lava called basaltic, which flows freely, the lava spreads out over a large area as it solidifies. When this is the case, the area around the vent has a gentle slope, and the volcano is known as a shield volcano—the shape resembles a broad shield, as illustrated in the figure on page 70. In the opposite case, involving lava known as rhyolite or felsic that is so thick it hardly flows at all, the lava builds up around the vent because it does not travel far before it solidifies. This produces a volcanic dome, as can be seen in the figure below, where the area around the vent forms steep, rounded sides.

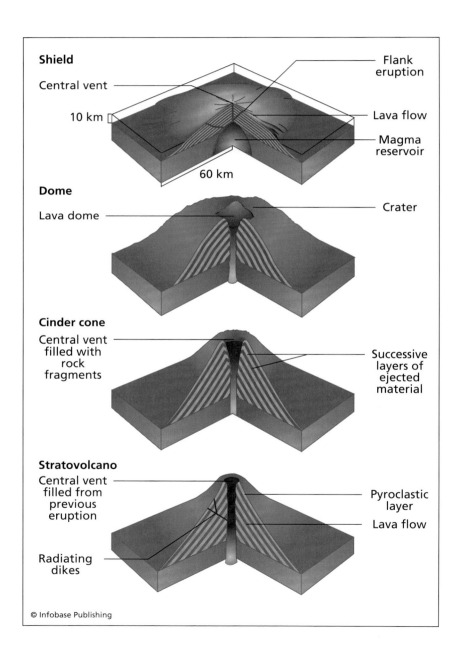

Shield

Central vent

10 km

60 km

Flank eruption

Lava flow

Magma reservoir

Dome

Lava dome

Crater

Cinder cone

Central vent filled with rock fragments

Successive layers of ejected material

Stratovolcano

Central vent filled from previous eruption

Radiating dikes

Pyroclastic layer

Lava flow

Other types of volcanoes emit pyroclastic flows—these particularly dangerous emissions include hot ash, gases, and dust that form a huge rolling cloud of material. (The term *pyroclastic* comes from Greek words *pyr,* meaning "fire," and *klastos,* meaning "broken" or "fragmented.") The cloud travels quickly, trapping unwary people in its path. The Mount Pelée eruption in 1902 included a pyroclastic flow, which was one of the reasons why it was so deadly. Volcanoes that emit pyroclastic flows tend to have the shape of a cinder cone, as illustrated in the figure on page 70. (Movies and television shows often portray volcanoes having this particular shape.) Some volcanoes, called stratovolcanoes, may emit lava or pyroclastic flows, which build up into a cone shape having alternating layers. Examples of stratovolcanoes include Vesuvius, Etna, and Mount Rainier in Washington.

A *caldera* is a basin or sink formed by the collapse of a volcanic structure, which can occur after an eruption of a large magma chamber below the vent empties and is unable to support the overlying earth. The Yellowstone volcano is a caldera. This caldera is impressive in size, covering an area larger than the state of Rhode Island.

Eruptions do not flow from a central vent in some volcanoes, but instead arise from fissures or long cracks in the surface. Fissure eruptions can release huge quantities of material. The mid-ocean ridges, discussed in chapter 2, are the sites of many fissure eruptions.

Magma is hot, which makes it less dense than surrounding rock, and so it rises, as does air heated over a stove. Earth's interior is clearly hot enough to contain melted rocks, and the asthenosphere is the primary source of magma. Deep below the surface, magma may rise through cracks or force its way upward by melting the rocks above it, forming the large chambers that fuel volcanoes.

Eruptions can be gentle outpourings, or they can be violent events. Scientists do not fully understand the processes that cause eruptions to occur at specific times, although heat, pressure, and plate movements are important factors. In many active volcanoes, lava from previous eruptions solidifies above the vent, temporarily plugging the opening.

(opposite page) Volcano types include shield, dome, cinder cone, and stratovolcano.

Strokkur Geysir in Iceland
*(Lukáš Hejtman/
iStockphoto)*

As the pressure rises, the volcano "blows its top" with a tremendous explosion. Aiding these explosions is the presence of certain gases, primarily water vapor along with carbon dioxide and sulfur dioxide. The heat will turn the water vapor into hot, high-pressure steam, which seeks an escape valve—and this steam often has the power to create one, if the path is blocked. Sometimes these explosions can be cataclysmic; a volcanic eruption destroyed the Indonesian island of Krakatoa in 1883, creating deadly tsunamis and a noise that people reported hearing all the way in Australia, 2,170 miles (3,500 km) away!

Volcanic activity need not involve outpouring lava or pyroclastic flows. Water is usually present in the ground, seeping through the soil after a heavy rain, and if such groundwater meets magma, it will become heated and rise. Some of this hot water or steam travels all the way to the surface, where it becomes the basis of hot springs and geysers, which are hot-water jets that periodically spout from the surface. Yellowstone National Park contains many of these springs and geysers, such as Old Faithful, a geyser that erupts every 65 minutes. Such activity suggests that the Yellowstone volcano is not quite deceased.

Large-scale eruptions such as the 1883 Krakatoa event have serious global implications. An eruption in 1815 of another Indonesian volcano, Mount Tambora, spewed enough gas and ash into the atmosphere to change the weather for a year. Before finally dispersing, the atmospheric contaminants blocked or absorbed sunlight and resulted in an exceptionally cool summer, including snow in New England in July of 1816! Eruptions can even lead to extinction of a considerable number of species if the changes are severe and last a long time. Some paleontologists—scientists who study ancient life—believe massive volcano eruptions may have caused or contributed to some of the mass extinctions that have occurred in the past, as observed in the fossil record. For instance, many species became extinct about 225 million years ago, and volcanic eruptions, along with their aftereffects, may have been involved. (Many people believe the extinction of the dinosaurs, which took place about 65 million years ago, was due to a comet or asteroid impact, but volcanic eruptions may have also played a role.)

The heat of Earth's interior explains the origin of the molten rock, steam, and other hot material that erupts from vents and fissures on the surface of the planet, but this explanation does not help geologists understand why volcanoes form or why a volcano arises at one location and not another. Asking this sort of question led scientists to map the distribution of active volcanoes, specifying the position of each one. The resulting map shown on page 74 provides a simple and practical reason for the location of the majority of Earth's volcanoes—and poses an intriguing puzzle for the remaining volcanoes.

RING OF FIRE

Geologists have determined that Earth's lithosphere (crust and uppermost mantle) is composed of slabs of rock called tectonic plates (see chapter 1). There are about a dozen major plates and a few dozen smaller ones. The thickness of these plates averages about 60 miles (100 km), but the crust underneath the ocean is thinner than the crust of the continents. Because of the fluidity of Earth's mantle, the plates do not stay in one place but slowly shift positions. The boundary between two plates is a site of much activity, as the moving plates collide, grind past one another, or diverge (move apart). Earthquakes and volcanoes frequently arise at plate boundaries.

A map of active volcanoes around the Pacific Ocean (see page 74) shows that the location of volcanoes in this region is not at all random. Volcanoes congregate along the boundary of the Pacific plate and other tectonic plates. Mount St. Helens, Krakatoa, and many other volcanoes form a ring of fire encircling the Pacific Ocean.

Most of Earth's volcanoes—about 95 percent—occur at the borders between tectonic plates. Magma surges through the cracks or seams, providing the material for volcanic eruptions. At some boundaries,

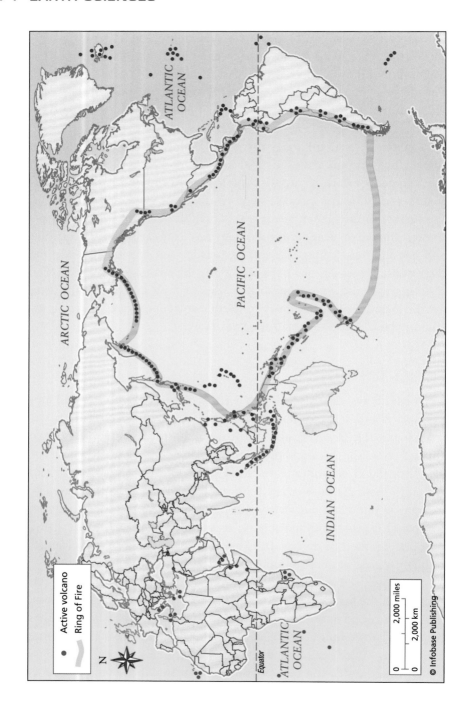

(opposite page) Numerous volcanoes, shown as dots, line up at the plate boundaries around the Pacific Ocean.

where plates collide and one plate dives beneath another, mountainous volcanoes form along the upper plate. For instance, Vesuvius is located at a plate boundary in which the African plate appears to be sliding underneath the Eurasian plate (but the geology of this region is complicated). Diverging plates, as occur along the Mid-Atlantic Ridge, create long valleys or fissures from which a huge quantity of lava extrudes.

Explaining the position of 95 percent of the world's volcanoes is an important achievement. With one theory—plate tectonics—the location of the vast majority of Earth's volcanoes becomes understandable. But what about the remaining volcanoes? Kilauea in Hawaii, for example, is far from a plate boundary. Yellowstone is another volcano situated away from plate boundaries. The location of these and other such volcanoes forces geologists to search for another explanation.

HOT SPOTS

One theory to explain Kilauea and other such volcanoes emerged in 1963. The Canadian geologist J. Tuzo Wilson (1908–93) observed that Hawaiian volcanoes have been pumping out a lot of lava that solidified a long time ago. The Hawaiian Islands are a chain of islands located in a small region in the Pacific Ocean. The main islands are, from west to east, Kauai, Oahu, Maui, and Hawaii. (The island chain is named after Hawaii, which is also known as the Big Island, as it is the largest in size though not in population—Oahu, home of the capital, Honolulu, has the most people.) Wilson proposed the existence of a small, stable, hot region—a hot spot— to account for the presence of these volcanic islands.

Movement of the tectonic plates enters into Wilson's theory if a hot spot is stable and located at a depth beneath the crust. If a hot spot does not move with the overlying tectonic plate, then as the plate slides along, the volcanic activity will appear to move in the opposite direction, as shown in the figure on page 77. This is because the hot spot is

Molokini, a volcanic crater that forms a crescent-shaped island near Maui, Hawaii *(Ron Chapple/Getty Images)*

steady, entrenched in the mantle, and the plate will move past it. Plates move slowly—in a range of one to six inches (2.5–15 cm) a year—but over millions of years, this movement is significant. Wilson's theory suggested that the chain of Hawaiian islands were formed from this volcanic activity in a manner that reflects the movement of the plate. As the plate lingers over the hot spot, the volcanic activity builds a sea-mount, which gradually rises above the surface to create an island. After some period of time, as the plate moves on, another island in the chain forms, slightly behind the earlier island. Wilson published his theory in a report, "A Possible Origin of the Hawaiian Islands," in a 1963 issue of the *Canadian Journal of Physics*.

Kilauea is a highly active volcano located on the island of Hawaii (the Big Island). According to Wilson's theory, the islands farthest from the current hot spot should be the oldest, since they were formed much earlier. Kauai, the most northwestern of the major islands in the chain, has rocks as old as 5 million years. This age contrasts with the Big Is-land—the most southeastern island—in which the oldest known rocks are less than 1 million years. The ages of the other islands also agree with the theory.

Hawaiian volcanoes have been extremely important not only for volcanologists interested in hot spot theory, but also for legions of tourists and interested onlookers. Native islanders have been observing these volcanoes for many generations, and the British explorer Captain James Cook (1728–79) sighted the Hawaiian Islands in 1778. Written records of Kilauea began in 1790, showing that the volcano has been active for most of the past two centuries. In periods of high activity, such as during eruptions or when lava rises to a visible level, Kilauea draws a crowd. People such as Mark Twain (1835–1910), who visited the volcano in 1866, began writing about their experiences, and the rest of the world became aware of the fascinating spectacle. In 1916 the U.S. government established Hawaii Volcanoes National Park, which includes Kilauea on the Big Island. Hawaiian volcanoes continue to be

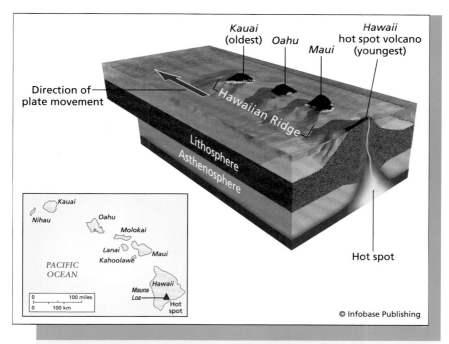

© Infobase Publishing

As the plate moves over the hot spot, a series of volcanoes form.

the site of important observations and studies, as described in the following sidebar.

The Hawaiian Islands are the youngest in an extended chain of volcanic islands and undersea mountains (which do not quite reach the surface) stretching about 3,700 miles (5,920 km) across the Pacific Ocean. This chain is known as the Hawaiian-Emperor Seamounts. Ages of the rocks indicate a progressive increase from northwest to southeast—the oldest rocks in the northwestern islands and seamounts are millions of years older than those of the southeast, and the age increases the farther one moves to the northwest. This "trail" probably marks the track of the Pacific plate's motion, but as shown in the following figure,

Hawaiian Volcano Observatory

Scientists who are seeking active volcanoes have found Kilauea extremely attractive. Perret, the pioneering volcanologist, visited Kilauea in 1911, and a year later the Massachusetts Institute of Technology professor Thomas A. Jaggar (1871–1953) began excavating along the rim of the Kilauea caldera. Jaggar and his team built a structure with a cellar that housed a seismometer, which he used to monitor the activity of the region. Money for this kind of geological research became easier to obtain after the disaster in Martinique in 1902 and the devastating earthquake in San Francisco in 1906, as people started to realize the value of volcanic and seismic research for society as well as science. The facility at Kilauea was the beginning of the Hawaiian Volcano Observatory.

Today the Hawaiian Volcano Observatory is a component of the Volcano Hazards Program of the United States Geological Survey (USGS). (The history and functions of USGS are outlined in a sidebar on page 10.) Researchers at the observatory study Kilauea and Mauna Loa, another volcano there is a sharp bend at about the middle of the chain that is not yet fully understood, corresponding to about 42 to 48 million years ago. The plate may have changed direction at this point, or the hot spot may not be stationary, as discussed below.

Chemical analysis of the lava from the Hawaiian volcanoes yields clues about their origin. The lithosphere is about 50 miles (80 km) thick underneath the Hawaiian Islands, so if the magma is coming from underneath, it might be of a different chemical nature than the lava erupting from shallow mid-ocean ridges. In particular, geologists have examined the ratio of isotopes of certain elements such as helium.

on the Big Island. Mauna Loa is an active volcano, erupting more than 30 times since 1843, although it has not erupted since 1984. This massive shield volcano is the largest volcano on Earth—the mountain covers about half of the island and rises 2.4 miles (4 km) above sea level; its flanks extend another three miles (five km) beneath the surface of the ocean.

It was the robust activity of these volcanoes that drew geologists to the site, and researchers at Hawaiian Volcano Observatory continue to monitor and track the volcanoes' behavior, study the history of their eruptions by analyzing volcanic rocks in the area, and inform the public of the nature and potential hazards of these geological phenomena. In addition, because the Hawaiian volcanoes are not on a plate boundary, these volcanoes are important testing grounds for hot spot theories, although researchers did not know of this benefit when they initially set up the observatory. Scientific advances come about because of the persistence, intelligence, and, occasionally, good fortune of scientists. Researchers who explore the frontiers of knowledge never know in advance exactly where a project will take them or how rewarding it will be.

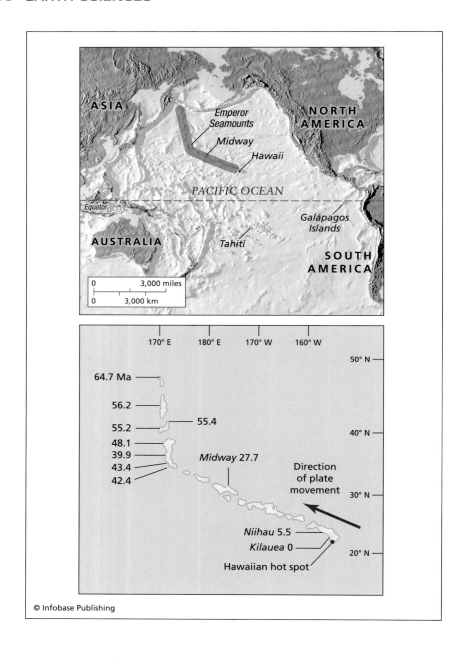

Atoms of the same element may have a different number of neutrons in their nucleus, resulting in different isotopes such as helium-3 (which has three particles in the nucleus, two protons and one neutron) and helium-4 (which has two protons and two neutrons in the *(opposite page)* The volcano trail that apparently tracks the movement of the plate makes a sharp bend about 42 to 48 million years ago.

nucleus). These isotopes usually have the same chemical properties but differ in stability—some isotopes are highly radioactive, decaying into other nuclei by emitting certain particles. Helium-3 and helium-4 are both stable, but helium-4 is a product of a number of different radioactive decays and is far more abundant. The ratio of helium-3 to helium-4 varies from place to place, and although this ratio is sometimes difficult to measure—helium is a highly mobile element—it does get trapped in Earth's crust. Deeper sources tend to have more helium-3, which was left over from Earth's formation as it coalesced from dust and gas in the galaxy. (Helium and most other elements are made in stars, which form the materials of Earth and living creatures.)

The helium-3/helium-4 ratio is distinctly higher in volcanic rocks from Hawaii than in the volcanic rocks of the mid-ocean ridges. This difference suggests, although it does not prove, a different origin for the magma of these two volcanic systems, with the Hawaiian system being fed by deeper sources. Other isotope ratios from these two systems also have differing values.

Hawaii is not the only hot spot in the world. Recall that about 5 percent of the world's active volcanoes are found at significant distances from plate boundaries. Yellowstone is another example of hot spot volcanic activity. As in the Hawaiian-Emperor seamount chain, a track of past volcanism marks a path from southern Oregon through Idaho and on into Wyoming and Yellowstone, which may indicate the movement of the North American plate as it glides past the hot spot.

The existence of volcanoes far from plate boundaries compels geologists to accept an alternative explanation for these volcanoes. In some form or fashion, magma travels through the middle of a plate. But the properties and origin of this magma have not yet been determined. One idea involves narrow channels called *plumes* reaching as far down as the deepest part of the mantle, 1,800 miles (2,900 km) beneath Earth's surface.

PLUME HYPOTHESIS

The Princeton geologist W. Jason Morgan published a paper, "Convection Plumes in the Lower Mantle," in a 1971 issue of *Nature*. Morgan extended Wilson's hot spot idea by proposing the existence of deep channels called plumes in which hot materials flow and transfer heat by the mechanism of convection currents: "In my model there are about twenty deep mantle plumes bringing heat and relatively primordial material up to the asthenosphere and horizontal currents in the asthenosphere flow radially away from each of these plumes." As tectonic plates move over these magma jets, the molten rock burns a hole through the plate, forming a hot spot volcano. This gives magma a channel to the surface without having to seep through cracks between tectonic places. Morgan believed these plumes play a role in continental drift.

The plume hypothesis is a simple idea. When approaching a problem, scientists usually consider the simplest solution first—it is the easiest one to test, and there is no reason to make the situation any more complicated than necessary. Geologists would love to be able to pry open the surface beneath these hot spots and check for any plumes. Finding such plumes would be a powerful piece of evidence supporting the hypothesis.

Searching the crust and mantle by drilling deeply into the surface is not possible at the moment, so the "eyes" geologists use to study Earth's interior are seismic waves. By accumulating enough seismic wave data, scientists can generate a three-dimensional image of the planet. Seismic tomography is the name of the technique geologists use to generate these images. The term *tomography* comes from Greek words *tomos,* meaning "section," and *graphein*, "to write"; tomography is the process of combining sections or slices of data into a three-dimensional image.

A geologist's use of seismic tomography is similar to a physician using ultrasound waves to map the interior of a patient's body. But the great size of the planet obscures images of the deepest parts. Adequate pictures of the upper mantle are possible, and geologists have found narrow channels that may be plumes, including one under Hawaii. But no one is sure how deep these channels extend.

Recent improvements in seismic tomography have given geologists the opportunity to probe even deeper. Raffaella Montelli, a researcher at Princeton University in New Jersey, and her colleagues at University of California, San Diego, University of Colorado, and National Taiwan University use a method known as finite-frequency tomography that combines a larger number of data sets. The method increases resolution—the ability to discern small or narrow objects in an image. Montelli and her colleagues reported finding plumes that reach the lowest depths of the mantle in several hot spots, including the Pacific islands of Tahiti and Easter Island. The picture underneath Hawaii was fuzzier, but there may also be a similar plume there as well. These plumes range in diameter from 60–240 miles (100–400 km). Montelli and her colleagues published their report, "Finite-Frequency Tomography Reveals a Variety of Plumes in the Mantle," in a 2004 issue of *Science.*

Although the latest methods of seismic tomography provide some visual evidence for the existence of plumes, images of vast depth can be hazy and difficult to interpret. These channels may not be plumes at all. And tomography has not been able to find candidate channels in all hot spot regions.

If the plumes exist, how do they form? Scientists usually consider evidence for an object more compelling if there is a convincing explanation of how it can arise. If a plume is an improbable event that is difficult or even impossible to understand, geologists will be more inclined to look for alternatives to explain the channels seen with seismic tomography without having to resort to plumes.

The long trail of islands in the Hawaiian-Emperor seamount chain indicates that the hot spot currently under Hawaii has been active for millions of years. This means the plume, if there is one, must be quite stable. As mentioned earlier, the high temperatures and mobility of Earth's interior create convection currents that carry heat from the bottom upward, with the fluid rising as it gets hot and become less dense. Earth's mantle is rocky, but the heat is intense, especially in the lower depths, so the rocks are hot enough to undergo some degree of melting. This partial melting contributes to the shifting and gliding of the tectonic plates and may also provide an environment in which narrow jets—plumes—can survive for long periods of time.

Although the dynamics of Earth's interior may create opportunities for mantle plumes to form, how these narrow jets of magma actually arise is not at all obvious. But Anne Davaille, a researcher at the Institut de Physique du Globe de Paris (Institute of Geophysics of Paris) in France, along with other scientists, has experimented with miscible viscous fluids—substances that can flow and mix together. When different fluids are placed in contact, such as pouring milk in a cup of water, the fluids

Plumes and Superplumes

A stable plume is a narrow jet of flowing magma, but what a young plume may look like is subject to a great deal of debate. One scenario is that a plume begins at the boundary between the lower mantle and the liquid outer core, perhaps from a particularly violent wave or oscillation in the core. The plume may start out with a huge volume of molten rock flowing up through the mantle, followed by a more stable but thinner stream. This would give a plume an initial shape of a mushroom, with a broad top—the plume head—trailed by a narrow jet. When the plume head arrives at the surface, it would cover a broad area with magma, which would cool into igneous rocks. Such events may be responsible for broad plains of volcanic rock that geologists refer to as large igneous provinces.

In 1991 the University of Rhode Island researcher Roger Larson suggested that even greater events have occurred in Earth's history. Larson noticed that a huge swath of crust under the Pacific Ocean formed with extraordinary rapidity during part of the Cretaceous period, as determined by the age of these rocks. In the 40-million year span between 120 million and 80 million years ago, ocean crust production increased by about 1.5 times the normal rate, and there was a peak in the first 20 million years of this time frame.

will often mix and form a single combined substance rather than forming separate layers. But mixing does not always occur—oil and water, for example, do not mix—and the degree of blending depends on the properties of the fluids, such as thickness or resistance to flowing (viscosity).

Davaille has tested the behavior of fluids in a two-layer system, in which she adjusted the viscosity of the fluids by dissolving some amount of salt or cellulose in them. She heated the bottom layer and cooled the top layer, setting up a temperature gradient—a difference—between the

Larson referred to this dramatic increase during a short period of time (geologically speaking) as a pulse, as opposed to a steady formation. He proposed that a large plume event—a superplume—erupted underneath the Pacific Ocean basin. Larson published this idea in a paper, "Latest Pulse of Earth: Evidence for a Mid-Cretaceous Superplume," in a 1991 issue of *Geology*.

An interesting possibility associated with this Cretaceous superplume is the unusual lack of Earth's magnetic field reversals during this period. As discussed in chapter 2, the north and south poles of Earth's magnetic field have switched at random intervals, every 500,000 years on average. But the Cretaceous period contains a long stretch of time without such a reversal. The superplume and the stability of Earth's magnetic poles may be related in some way, although no one yet knows how or why.

Earth is not the only planet in the solar system with significant volcanic activity. On Mars, the Tharsis region is an elevated plateau about six miles (10 km) above the average surface level and covers about one-fourth of the planet's surface. Several large volcanoes dot this plain, including a shield volcano called Olympus Mons, which stands about 15 miles (24 km) high and is the largest known volcano in the solar system. Tharsis may be the result of a superplume, although this is only a speculative hypothesis. Martian geology will remain mysterious until the planet is more fully explored.

two layers. Convection occurred. But there was a lot of variability at the boundary between the layers, producing flows that resembled sheets or conduits rather than a broad mingling of the two fluids. Davaille reported this experiment, "Two-Layer Thermal Convection in Miscible Viscous Fluids," in a 1999 issue of *Journal of Fluid Mechanics*. This finding supports the notion that plumes can form in Earth's interior layers.

Laura E. Schmidt and Wendy W. Zhang, researchers at the University of Chicago in Illinois, analyzed and extended Davaille's findings. Schmidt and Zhang conducted a similar experiment with two layers of miscible oils and discovered convection that consisted of thin tendrils rising from the bottom layer to the top one. These tendrils resemble hypothetical mantle plumes. The researchers formulated a mathematical model of this phenomena, using mathematical equations describing fluid flow and dynamics. In some cases, the narrow convection tendrils can be stable for long periods of time, anchoring themselves at certain locations. Schmidt and Zhang published their report, "Viscous Withdrawal of Miscible Liquid Layers," in a 2008 issue of *Physical Review Letters.*

Narrow jets of magma are not the only possible convection patterns. Researchers have also considered the possibility of temporary but massive upwellings of magma, rising to the surface and spilling a lot of material at once. Sometimes these events are called superplumes, as described in the sidebar on page 84.

But plumes are not the only explanations for hot spot volcanism. Although sophisticated techniques such as seismic tomography present some evidence that plumes exist, the data is sketchy. While geologists continue to collect data, observations that are not in accordance with the plume hypothesis have also emerged.

ALTERNATIVE HYPOTHESES TO EXPLAIN HOT SPOTS

Marcia K. McNutt, a scientist at the Monterey Bay Aquarium Research Institute in California, made the following observation in the September 8, 2006, issue of *Science*: "Scientists love beautiful theories—the kind that are elegant, predictive, and have few free parameters. And they hate it when theories like that prove to be wrong. It is thus with much kicking, dragging, and screaming that geoscientists are being brought to the realization that all might not be well with the concept of mantle plumes."

One of the agitating factors McNutt referred to is the recent findings of Naoto Hirano of the Tokyo Institute of Technology in Japan. Hirano, along with a team of other researchers, recently found young volcanic rocks in the Pacific Ocean near the coast of Japan. These volcanic seamounts are perched on an old section of the Pacific plate, at some distance from any boundary. This location puts these volcanoes in the hot spot category. But the researchers conducted chemical analyses on rock samples and found isotope ratios similar to those found in mid-ocean ridges, suggesting a more shallow source for this material than other hot spot volcanoes. This finding argues against the existence of a deep mantle plume here. Another problem with the plumes hypothesis in this case is that an earlier tomography survey indicated no trace of a channel or conduit beneath this area. As McNutt wrote, "The authors describe a small chain of hot spot volcanoes off the Japanese coast that almost assuredly cannot have been formed by narrow, deep-Earth upwellings."

As an alternative explanation, Hirano and his colleagues proposed that the magma originated in the asthenosphere. The eruptions may have occurred because the Pacific plate flexed or cracked at this point, possibly due to its collision farther west with the Eurasian plate, beneath which the Pacific plate is diving. These cracks would not be very deep, but they could allow shallow magma to rise and erupt. Hirano and his colleagues published their report, "Volcanism in Response to Plate Flexure," in a 2006 issue of *Science.*

The results of Hirano and his colleagues do not disprove the plume hypothesis, although their research does provide an example of hot spot volcanic activity that does not appear to be caused by deep mantle plumes. Shallow pools of magma may also fuel other hot spot volcanoes, although the isotope ratios of Hawaii and many of the others in this category point toward a deeper source. As geologists probe deeper into Earth—including the ambitious attempts to drill into the mantle, as described in chapter 1—a better understanding of the chemistry of Earth's interior will emerge. Such knowledge will shed much light on these questions.

Scientists will also continue to develop elaborate models to help them explore and understand the hidden processes occurring in Earth's interior. To study the structure and kinematics—motion—of the Pacific plate, Valérie Clouard and Muriel Gerbault of the University of Chile devised a mathematical model of this plate. The researchers focused on the forces acting on the plate as it moves across the surface of the planet, colliding with other plates. These collisions, along with volcanic activity at the plate boundaries, generate enormous stresses and strains, resulting in some amount of deformation of even a rigid, rocky plate.

Clouard and Gerbault were particularly interested in studying the behavior exhibited by their model around an area of the central Pacific Ocean. This region contains several hot spots—Samoa, Cook Islands,

Austral Islands, Tahiti, Marquesas Islands, and Pitcairn (the island to which Fletcher Christian and the other mutineers of the HMS *Bounty* fled). The researchers performed a simulation of the Pacific plate's movement over the last few million years and discovered a shearing force—a force tending to twist or tear—occurring around the middle of the plate. Cracks opening up in this area could explain the magma upwellings without the need of a deep mantle plume—magma from more shallow sources can rise through the fracture. Clouard and Gerbault published their report, "Break-up Spots: Could the Pacific Open as a Consequence of Plate Kinematics?," in a 2008 issue of *Earth and Planetary Science Letters*. From their results, the researchers concluded that "the Pacific intraplate volcanism would correspond to the formation of melting columns in the upper asthenosphere, in response to shearing plate boundary conditions. Central Pacific hot spots should be seen as break-up spots of shallow origin." This suggestion is similar to the idea put forward by Hirano and his colleagues. Clouard and Gerbault also suggested that if the tearing force continues, the Pacific plate may be on its way to splitting, although this would require a great deal of time.

These simulations and models of Clouard and Gerbault suggest an alternative to the plume hypothesis for hot spots such as Samoa, even though the seismic tomography study described earlier offers evidence for a deep mantle plume in this region. Competing ideas are productive in science because they spur scientists to collect more data in order to decide which of the competitors, if any, is correct. The plume hypothesis and its alternatives are controversial at the present time and will remain so until the issue is resolved. Perhaps the solution will involve all of the these ideas, in one form or another, at various spots on the planet—Earth is a large place and not generally uniform, so there is room for more than one mechanism or process associated with hot spot volcanoes.

Another interesting question raised by the work of Clouard and Gerbault, along with other researchers, concerns the mobility of hot spots. Some scientists who have tracked the path of hot spot volcanoes such as the Hawaiian-Emperor seamount chain have assumed hot spots are stationary. Yet if plate movements and stresses are the source of the volcanic activity, the hot spots may shift around depending on the forces acting on the plate.

Anthony A. P. Koppers and Hubert Staudigel of Scripps Institution of Oceanography in California studied bends in the Gilbert ridge and Tokelau seamounts in the Pacific Ocean. These bends are similar in angle to the bend in the Hawaiian-Emperor seamount chain (see the figure on page 80). If all of these bends are due to changes in the direction of motion of the Pacific plate, as described above, then they should have occurred at about the same time. As mentioned earlier, geologists believe the bend in the Hawaiian-Emperor seamount chain occurred around 42 to 48 million years ago.

However, when Koppers and Staudigel dated samples of the rocks they obtained at the bends of the Gilbert ridge and Tokelau seamounts, they discovered that the bend in the Gilbert ridge may have occurred much earlier—about 67 million years ago—than that of the Hawaiian-Emperor seamount chain. They also found that the bend in the Tokelau seamounts may have happened earlier than previously believed, around 57 million years ago. The important thing is that the result suggests that the bends were asynchronous—occurring at different times—which is difficult to reconcile with the notion of a change in plate direction. In a rigid plate, a change in direction would occur throughout the plate at the same time. Koppers and Staudigel proposed that at least some of the hot spots in the Pacific Ocean are due to extensions of magma through local fractures rather than deep mantle plumes. The researchers published their findings, "Asynchronous Bends in Pacific Seamount Trails: A Case for Extensional Volcanism?," in a 2005 issue of *Science.*

CONCLUSION

In the nearly two millennia since the Vesuvius eruption of 79 C.E., scientists have learned much about volcanoes. The majority of volcanoes dot the boundaries between the enormous tectonic plates, where hot magma rises up through the seams of Earth's crust to fuel volcanic activity. What causes hot spots—and the small number of volcanoes situated away from plate boundaries—is not yet determined. In some cases, magma may spurt from the depths of the lower mantle in narrow plumes; in other cases, stress fractures in the middle of a plate may provide vents for shallow pockets of localized magma. Some hot spot volcanism may be due to some combination of the two, or some other, as yet unknown, mechanism.

In order to achieve a better understanding of hot spots and their causes, geologists will continue to study these phenomena. The determination to

Geologist measuring the height of a 1983 eruption of Kilauea volcano (J. D. Griggs/USGS)

learn more has not changed since ancient times, when Pliny the Elder risked his life to make close observations of a volcanic eruption. And the need for knowledge is greater than ever before. Millions of people live near active volcanoes, including hot spot volcanoes. A more complete understanding of these volcanoes is necessary before scientists will be able to predict their eruptions. Prediction is vitally important because it will give people in the path of danger ample time to escape, resulting in fewer casualties when cataclysmic eruptions occur.

Geologists can often detect the signals of an imminent eruption. The volcano may actually swell as it fills with magma, and the activity generates small earthquakes that are recorded on seismometers. For example, after more than a century of dormancy, Mount St. Helens suddenly began experiencing small tremors in March 1980. This activity prompted USGS to issue a hazard alert. As the seismic waves increased in intensity over the next month, USGS officials raised a more serious alarm, and most of the people living in the danger zone evacuated the site. These actions greatly decreased the loss of life, but no one could tell exactly when, or if, the volcano would erupt. On May 18, 1980, Mount St. Helens finally erupted with an explosion that sheared off about 1,300 feet (400 m) of the mountain.

Improved warning systems will save even more lives and decrease the disruption of lengthy or unnecessary evacuations. The latest developments use advanced technology such as satellite imagery. For example, Falk Amelung, a geologist at the University of Miami in Florida, along with colleagues in Miami and at Stanford University in California, recently used the sophisticated radar of a Canadian satellite to map the area around Mauna Loa. The radar measures the distance between the ground and the satellite with a high degree of accuracy, providing researchers with a precise measurement of ground deformation that is associated with magma flow and volcanic activity.

Mauna Loa, the world's largest single volcano, poses a serious threat to Hawaiian residents, including the potential danger of a tsunami. These high-speed waves flood coastal areas, as in the Indian Ocean tsunami of 2004 that claimed hundreds of thousands of lives. Triggers for these disastrous waves include undersea earthquakes and landslides falling into the ocean, such as might occur if a gigantic eruption collapses a wall or flank of Mauna Loa.

The satellite images gave Amelung and his colleagues the ability to observe magma flows along rift zones—narrow valleys or cracks where lava may extrude. As more magma enters the cracks, the stress of this additional material pushes the crack wider, possibly leading to a significant eruption. By tracking deformations with radar, geologists may be able to get a better idea of when and where an eruption is likely to occur. Amelung and his colleagues published their findings, "Stress Control of Deep Rift Intrusion at Mauna Loa Volcano, Hawaii," in a 2007 issue of *Science.* The researchers showed that "the stress field within the volcanic edifice is a dominant effect in controlling magma accumulation. Space-geodetic measurements can be used to infer changes to the stress field in the interior and contribute to better forecasts of the response of a volcano to the arrival of new magma from below."

Another potentially dangerous hot spot is Yellowstone. With the aid of satellite radar and global positioning system (GPS) equipment, which uses satellites to pinpoint the coordinates of a given spot on Earth, Wu-Lung Chang, Robert B. Smith, and their colleagues at the University of Utah and USGS observed an increased rate of uplift in the Yellowstone caldera. The accelerated uplift, which occurred during the years 2004 to 2006, suggests a large quantity of magma is accumulating and expanding the chambers underneath the surface. This episode does not imply that an eruption or any elevated volcanic activity is forthcoming, but the size of this system is so large that geologists will keep a close eye on it. The researchers published this discovery in a paper, "Accelerated Uplift and Magmatic Intrusion of the Yellowstone Caldera, 2004 to 2006," in a 2007 issue of *Science.*

A Yellowstone eruption would have a huge impact on the United States, and, as gigantic eruptions of the past have shown, could affect global weather patterns as well. The risk at present seems low, but a more accurate theory of hot spots would lead to more confident evaluations. Volcanoes and hot spots act as windows to Earth's fiery interior, possibly as far down as the lowest depths of the mantle, and are at the edge of a frontier of science that is essential to understand to predict some of the most important hazards facing the world today.

CHRONOLOGY

79 C.E.	Mount Vesuvius in Italy erupts, burying the city of Pompeii in ash and lava and destroying other towns and houses nearby.
1815	Tambora in Indonesia erupts, sending so much gas and ash into the atmosphere that Earth's temperature was temporarily cooled, resulting in a snowy summer in New England and elsewhere the following year.
1847	Italian scientists establish a volcanic observatory at Vesuvius.
1883	The volcanic Indonesian island of Krakatoa explodes in an eruption that was one of the most violent events in history.
1902	Mount Pelée, a volcano on the Caribbean island of Martinique, erupts, destroying the city of Saint-Pierre and killing 30,000 people.
1912	The Massachusetts Institute of Technology professor Thomas A. Jaggar (1871–1953) and his colleagues begin constructing what will become the Hawaiian Volcano Observatory.
	The German researcher Alfred Wegener (1880–1930) proposes that Earth's continents drift over time. Although the idea was incorrect in some of its

details, scientists later develop the theory of plate tectonics, in which pieces of Earth's crust move slowly. Maps of volcano sites show that 95 percent of the world's active volcanoes are at or near a plate boundary, at which point plates collide, separate, or grind past one another.

1963   The Canadian geologist J. Tuzo Wilson (1908–93) proposes the existence of a hot spot—a small region where magma rises to the surface through a channel—to account for the presence of volcanoes far removed from tectonic plate boundaries.

1971   The Princeton geologist W. Jason Morgan publishes a paper, "Convection Plumes in the Lower Mantle," in which he explains hot spots by the presence of deep channels through the mantle called plumes.

1980   Mount St. Helens in Washington erupts, killing 57 people.

1991   Roger Larson proposes that a large plume—a superplume—is responsible for the outburst of crust formation under the Pacific between 120 million and 80 million years ago, during the Cretaceous period. This event is sometimes called the Cretaceous superplume.

2004   Using advanced seismic tomography, Raffaella Montelli and her colleagues find evidence for deep mantle plumes under several hot spots.

2006   Naoto Hirano and his colleagues report their investigation of young volcanic seamounts, perched on an old section of the Pacific plate at some distance from any boundary. But the researchers found evidence supporting a shallow rather than a deep source for the volcanic material, contrary to the plume hypothesis.

FURTHER RESOURCES

Print and Internet

Amelung, Falk, Sang-Ho Yun, et al. "Stress Control of Deep Rift Intrusion at Mauna Loa Volcano, Hawaii." *Science* 316 (May 18, 2007): 1,026–1,030. Amelung and his colleagues use satellite imagery to study magma flows along rift zones.

De Boer, Jelle Zeilinga, and Donald Theodore Sanders. *Volcanoes in Human History: The Far-Reaching Effects of Major Eruptions.* Princeton, N.J.: Princeton University Press, 2002. This book covers famous and disastrous volcanic eruptions, including Vesuvius in 79 C.E., Tambora in 1815, Krakatoa in 1883, Mount Pelée in 1902, Mount St. Helens in 1980, and others.

Chang, Wu-Lung, Robert B. Smith, et al. "Accelerated Uplift and Magmatic Intrusion of the Yellowstone Caldera, 2004 to 2006." *Science* 318 (November 9, 2007): 952–956. The researchers detect an increased rate of uplift in the Yellowstone caldera during the years 2004 to 2006.

Clouard, Valérie, and Muriel Gerbault. "Break-up Spots: Could the Pacific Open as a Consequence of Plate Kinematics?" *Earth and Planetary Science Letters* 265 (2008): 195–208. This model suggests that some Pacific hot spots are due to intraplate cracks and shallow magma sources.

Davaille, Anne. "Two-Layer Thermal Convection in Miscible Viscous Fluids." *Journal of Fluid Mechanics* 379 (1999): 223–253. Davaille reports on tests of the behavior of fluids in a two-layer system that supports the notion of plumes.

Decker, Robert, and Barbara Decker. *Volcanoes,* 4th ed. New York: W. H. Freeman, 2005. Offering a wealth of information, Robert and Barbara Decker describe the properties of volcanoes and eruptions, with enough detail so that the reader learns not only about these properties but also how volcanologists go about studying them.

Fisher, Richard V., Grant Heiken, and Jeffrey B. Hulen. *Volcanoes.* Princeton, N.J.: Princeton University Press, 1998. Written by expert volcanologists, this book discusses topics including eruptions and why they occur, hazards such as lava flows and ash clouds, and the myths and allures of volcanoes.

Foulger, Gillian R. "Mantle Plumes." Available online. URL: http://www. mantleplumes.org/. Accessed May 4, 2009. Founded in 2003, Mantle-Plumes.org and the associated Web site aim to publicize the debate and discussion over the issue of the possible causes of hot spot volcanoes. The Web resource includes articles on Earth's mantle, plumes and superplumes, hot spots, and related subjects. Most of the articles are contributions of geologists and experts, and although some of these articles are written at an advanced level, all the fundamental problems and debates swirling around this issue are discussed.

Giblin, Mildred. "Frank Alvord Perret." *Bulletin of Volcanology* 10 (1950): 191–196. The article recaps Perret's life and scientific career.

Hirano, Naoto, Eiichi Takahashi, Junji Yamamoto, et al. "Volcanism in Response to Plate Flexure." *Science* 313 (September 8, 2006): 1,426–1,428. This paper describes evidence of a hot spot volcano system that appears not to be due to plumes.

Koppers, Anthony A. P., and Hubert Staudigel "Asynchronous Bends in Pacific Seamount Trails: A Case for Extensional Volcanism?" *Science* 307 (February 11, 2005): 904–907. The researchers propose that at least some of the hot spots in the Pacific Ocean are due to extensions of magma through local fractures rather than deep mantle plumes.

Krystek, Lee. "Is the Super Volcano Beneath Yellowstone Ready to Blow?" Available online. URL: http://www.unmuseum.org/ supervol.htm. Accessed May 4, 2009. Discussions of the giant Yellowstone volcano and its possible return to activity are usually accompanied by much hyperbole and as much heat and hot air as any volcanic eruption. This article, an entry in the Museum of UnNatural Mystery, offers some of the facts and some of the speculation.

Larson, R. L. "Latest Pulse of Earth: Evidence for a Mid-Cretaceous Superplume." *Geology* 19 (1991): 547–550. Larson proposes that a large plume event—a superplume—erupted underneath the Pacific Ocean basin.

McNutt, Marcia. "Another Nail in the Plume Coffin." *Science* 313 (September 8, 2006): 1,394–1,395. McNutt reviews evidence against the plume hypothesis.

Montelli, Rafaella, Guust Nolet, et al. "Finite-Frequency Tomography Reveals a Variety of Plumes in the Mantle." *Science* 303 (January 16,

2004): 338–343. Montelli and her colleagues report finding plumes that reach the lowest depths of the mantle in several hot spots.

Morgan, W. Jason. "Convection Plumes in the Lower Mantle." *Nature* 230 (March 5, 1971): 42–43. Morgan extends Wilson's hot spot idea by proposing the existence of deep channels called plumes in which hot materials flow and transfer heat.

Pliny the Younger. *The Letters of the Younger Pliny.* New York: Penguin Classics, 1963. This book reprints the letters of this ancient scholar.

Schmidt, Laura E., and Wendy W. Zhang. "Viscous Withdrawal of Miscible Liquid Layers." *Physical Review Letters* 100 (2008): 044502.1–044502.4. Available online. URL: http://arxiv.org/abs/0708.4293. Accessed May 4, 2009. The researchers formulated a mathematical model of mantle plumes.

Smithsonian Institution. "Global Volcanism Program." Available online. URL: http://www.volcano.si.edu/. Accessed May 4, 2009. Maps, reports, and photographs highlight this Web resource, which surveys large and small volcanic eruptions all over the world in the last 10,000 years.

Thompson, Dick. *Volcano Cowboys: The Rocky Evolution of a Dangerous Science.* New York: St. Martin's Press, 2000. How does a volcanologist study volcanoes? Very carefully, of course, but even so, it is a risky business. Thompson tells the story of volcanologists on the job at Mount St. Helens before and after the 1980 eruption and Mount Pinatubo in the Philippines, which erupted violently in 1991.

United States Geological Survey. "Hawaiian Volcano Observatory." Available online. URL: http://hvo.wr.usgs.gov/. Accessed May 4, 2009. The Web site of the Hawaiian Volcano Observatory provides a history of the observatory and the latest information on the status of the Kilauea and Mauna Loa volcanoes.

———. "'Hotspots': Mantle Thermal Plumes." Available online. URL: http://pubs.usgs.gov/gip/dynamic/hotspots.html. Accessed May 4, 2009. This brief article, although somewhat dated, is an excellent and accessible introduction to the subject.

Volcano World Team. "Volcano World." Available online. URL: http://volcano.oregonstate.edu/. Accessed May 4, 2009. An educational and public outreach project of the University of North Dakota and

Oregon State University, this Web resource is a great place for volcano enthusiasts to explore. Aimed at students as well as the general public, Volcano World contains a huge quantity of information and pictures of the world's volcanoes. The resource also includes interviews with volcanologists, updates on current eruptions, a glossary, and historical data.

Wilson, J. Tuzo. "A Possible Origin of the Hawaiian Islands." *Canadian Journal of Physics* 41 (1963): 863–870. Wilson describes his hot spot theory.

Winchester, Simon. *Krakatoa: The Day the World Exploded: August 27, 1883*. New York: HarperCollins, 2003. Although the subtitle is an exaggeration, the volcanic explosion of this Indonesian island was one of the most violent events in history. This book describes what researchers have learned about the explosion and discusses its scientific, geographical, and political aftereffects.

GEOTHERMAL ENERGY— A FURNACE BENEATH THE SOIL

The term *geothermal* comes from Greek words *geo,* meaning "Earth," and *therme,* meaning "heat." People have long observed geysers, hot springs, and volcanoes, indicating that the interior of the planet stores at least some amount of heat. The average temperature of Earth's crust increases about 72°F/mile (25°C/km), although there is considerable variability from place to place. For instance, hot spots and other sites of volcanic activity are considerably hotter at shallow depths than other areas. Earth's mantle also gets warmer with depth, although the temperature rise is not as great as the crust. Geologists can only estimate the core's temperature, but the outer core is probably 5,430°F (3,000°C) and the inner core may be as hot as 14,400°F (8,000°C).

Heat is energy—the ability to do work or make something move. Energy comes at a cost, such as the cost of food that people eat to provide energy (including body heat) or the cost of electricity or oil for the heater. Americans spend about $500 billion each year on energy, much of which involves heat. People use some of this heat to stay warm, but people also burn fuel in internal combustion engines, many of which use gasoline, and in the huge electric generators of the power companies, which run on high-pressure steam created by the heat from burning oil or coal. Most of the world's energy comes from burning these *fossil fuels,* which are hydrocarbons—substances consisting of compounds of hydrogen and carbon—

Pollution from smokestacks *(Rinderart/Dreamstime.com)*

including oil, natural gas, and coal. These substances are called fossil fuels because scientists believe they come from the remains of plants that lived and died long ago and were buried in sediments, where heat and pressure gradually transformed them into rich fuels. Although energy companies continue to find and extract fossil fuels from the ground, these substances are not a renewable resource, since there is only a limited and exhaustible supply. The limited supplies, coupled with the increasing demand of the world's growing population, have led to fuel shortages and spiking prices.

Burning fossil fuels is costly for the environment as well as the bank account. For example, hydrocarbon combustion produces pollutants responsible for smog. Yet about 85 percent of the energy in the United States in 2007 came from fossil fuels, according to estimates of the Department of Energy (DOE), the government agency responsible for advancing and developing energy technology.

Fossil fuel depletion, high prices, and environmental concerns motivate a search for clean, renewable sources of energy. The problem thus far has been cost, since fossil fuel alternatives tend to be more expensive, despite the rising cost of fossil fuels. However, one alternative lies beneath Earth's surface, in a vast reservoir of heat in the planet's interior. People have been using this energy source for a long time, albeit indirectly—the planet's heat and pressure was necessary to "cook" the fossil fuels that are now so widely exploited. But a more direct use of Earth's heat—geothermal energy—may abound in the future, if engineers and scientists can apply knowledge from the frontiers of Earth science to bring geothermal techniques into fruition. This chapter describes the successes that have been achieved and the research that aims to extend the use of geothermal energy even further.

INTRODUCTION

Earth's heat has two main sources. Part of the heat is left over from Earth's fiery creation, about 4.5 billion years ago. The other main contributor is radioactivity. Unlike the heat left over from Earth's formation, radioactive decay of certain isotopes within the planet is an ongoing process, adding more heat all the time. Some of the main sources of this heat are radioactive isotopes of uranium, thorium, and potassium.

Heat flows, or conducts, through objects—touching a hot stove is a bad idea because heat will flow from the stove to the skin, resulting in a burned finger. Another mechanism of heat transfer is a convection current. Convection currents are flows of air or liquid that carry heat and are important in many of geological processes described in the three previous chapters. Radiation is also an important heat transfer mechanism. All objects radiate, meaning that they emit electromagnetic radiation, which is a form of energy. The type and amount of radiation depends on the object's temperature. Hot objects emit radiation having a high frequency, such as visible light, which has more energy than low-frequency radiation such as infrared. Objects that are not as hot also emit radiation, but mostly infrared. This energy emission lowers the radiating object's temperature. The recipients of the emission—such as a sunbather on the beach who absorbs the Sun's electromagnetic radiation—get warmer.

Earth's internal heat and the planet's interactions with its surroundings govern its temperature, as is true for all objects. Conduction and convection carry heat from the hot interior to the cool surface; Earth's surface is cool because it radiates heat into space, mostly in the infrared portion of the electromagnetic spectrum. (Gases such as carbon dioxide and other greenhouse gases tend to absorb this radiation, a process that warms the planet. This effect is similar to what happens in a greenhouse, in which the panes of glass allow some of the sunlight to enter, but block infrared emissions of the objects within the house.) The surface also receives a great deal of heat as it absorbs some of the Sun's radiation. As a consequence of these interactions, Earth's temperature is relatively stable, although the planet was much warmer early in its history.

Parts of Earth's interior are hot enough to melt rocks, producing magma that fuels volcanoes, and to raise the temperature of water that seeps into the ground, producing geysers and hot springs. Ancient peoples took advantage of this heat by using springwater for bathing and cooking purposes. These springs were especially appreciated during cold winter months, such as those endured by inhabitants of northern Wyoming, the site of Yellowstone and its springs. Native Americans settled near and frequently visited most of the hot springs in Canada and the United States; archaeological artifacts show that people were using these sites as long as 10,000 or more years ago. Some people believe water from these hot springs possesses remarkable medicinal value, a belief that is commonly held today by patrons of spas located at various hot springs. Although curative properties of this springwater fall shy of being miraculous, the water often contains a great deal of minerals picked up as it traveled through the ground.

Ancient Romans were also avid users of springwater. Bathing was an important component of Roman society—citizens gathered at bathhouses to enjoy the water and discuss the latest news—and as a practical people, Romans took advantage of hot springs where they were available. Some of the Romans in the city of Pompeii, for example, used water from geothermal sources to heat their houses. Archaeologists made this discovery when they excavated Pompeii, which, as described in chapter 3, was buried by a volcanic eruption in 79 C.E. A portion of the city's buildings remain intact and have features such as plumbing to circulate hot water, allowing the heat to warm the interior.

EXPLOITING GEOTHERMAL ENERGY

The energy needs of modern times are much greater and more varied than those of ancient civilizations. Devices such as computers, engines, telephones, and many others require energy in order to function. People obtain much of this energy from the combustion of fossil fuels, which powers automobiles as well as electricity generators, but geothermal energy offers an alternative.

In addition to heating homes, as in Pompeii, heat from Earth's interior can be transformed into electricity. The Italian chemist and inventor Piero Ginori Conti (1865–1939); also prince of Trevignano, a *comune* or township in Italy) designed and built the first electric generator running from geothermal power in 1904. Working in Larderello in central Italy, where many hot springs are located, Conti used steam issuing from a well to drive a piston engine. The engine ran a dynamo, which is a device that generates electricity. It was a small experimental operation that had a meager output—it lit five lightbulbs, each of which consumed only a tiny amount of power—but the machine proved to be a success. Later, in 1911, the first geothermal power station appeared in Valle del Diavolo at Larderello.

Several dozen countries in the world today employ geothermal energy on a large scale. The list includes the United States. In addition to using geothermal energy for heating purposes, several states have built geothermal power stations to generate electricity. The majority of these stations are in California, which has more than 30 geothermal power stations that supply about 5 percent of the state's electricity. Nevada has more than a dozen geothermal power stations, located mostly in the northern section of the state. Alaska, Hawaii, and Utah have also built geothermal power stations. Although the total amount of electricity generated by geothermal power stations is small, their use saves Americans from paying for and burning millions of barrels of oil, millions of tons of coal, or large volumes of natural gas.

Geothermal power stations are similar to other types of electric generators. Most power stations in the United States and elsewhere generate electricity with giant turbines, which operate on the same principles of physics as a dynamo—a conductor spinning in a magnetic field produces electricity. A turbine is an engine consisting of a rotating shaft on

Iceland has abundant geysers and hot springs, such as those at Namaskard, near Lake Myvatn. *[Steve Allen/Getty]*

which blades are attached; high-velocity gas or liquid hits the blades, supplying the force of rotation. Power companies usually employ turbines to generate alternating current (AC), which is the type of electricity commonly used in appliances.

In a few power stations wind drives the turbine, and in other stations falling water supplies the energy, such as in the hydroelectric stations at Hoover Dam along the border of Nevada and Arizona in the United States. Most power companies use steam turbines, in which steam funneled at high pressure through the turbine presses against the turbine's blades, causing the shaft to rotate. In the majority of these power stations, the energy needed to boil water and create the high-pressure steam comes from burning fossil fuels. But as described in the following sidebar, geothermal energy offers an alternative.

Temperatures below the surface generally rise rapidly with depth, but some places are warmer than others, and geothermal steam or hot

Turning Geothermal Energy into Electricity

Electric generators are devices to convert energy in one form or another into electrical energy. Heat is a common form of energy that is transformed into electricity, as in turbines that are driven by a hot gas such as steam. The heat to create this hot gas can come from burning oil, coal, or natural gas, but it can also come from the Earth. Geothermal power stations use heat coming from beneath the surface to rotate the turbines.

There are three main types of geothermal power station, differing in the nature of the geothermal supply that they tap.

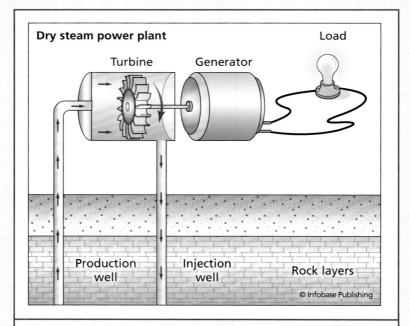

(A) In a dry steam power plant, the steam rises, turns turbine, then returns, in a cooler state, to the reservoir.

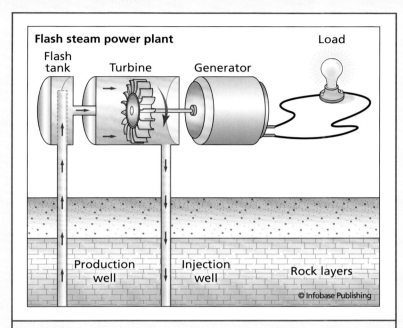

(B) In a flash steam power plant, hot water abruptly changes to steam in the flash tank due to the decreased pressure, then turns a turbine and returns to the reservoir.

A "dry" steam geothermal power station taps into a reservoir that is mostly vapor—steam—with little or no liquid (water), which is what gives it its name. Figure A on page 104 illustrates the basic operation. Pipes sunk into the underground reservoir bring steam into the turbine, where it rotates the shaft and drives the electric generator. The steam expends some of its energy in the turbine, which lowers its temperature. Pipes on the other side of the turbine return the fluid to the reservoir, so that it can be reheated and reused. Dry steam power stations are simple and were the first type of geothermal power station developed—the early genera-

(continues)

(continued)

tor of Italian inventor Conti was a rudimentary dry steam generator. The world's largest geothermal power station, 30 square miles (77 km²) along the Sonoma and Lake Count border, about 100 miles (160 km) north of San Francisco, California, is known as The Geysers. This dry steam power station harnesses naturally occurring steam field reservoirs below the Earth's surface.

Some of the reservoirs hold hot water instead of steam. These reservoirs can be used in a type of geothermal power station called a flash station or a flash steam station. Water deep below the surface can have a temperature in excess of the boiling point at sea level—212°F (100°C)—because the boiling point depends on pressure, and the high pressure beneath the surface means that water can exist at much higher temperatures without boiling. When this extremely hot water is brought to the surface and placed in an environment that does not exert as much pressure, the water rapidly boils or "flashes" into steam. As illustrated in figure (B), flash steam power stations employ this process to generate the steam needed to drive the turbine.

The third type of geothermal power station is called a binary station. This type of power station uses a geothermal reservoir containing water that is hot but not quite hot enough to operate a flash station. Instead, a piece of equipment called a heat exchanger transfers heat from the hot water to another fluid, which flashes at a lower temperature. This second fluid boils, producing the vapor that rotates the turbine. The term *binary,* referring to two components, comes from the use of two fluids.

water is not always readily accessible. In some of the western states of the United States, such as California and Nevada, geothermal reservoirs are within reach or in some cases rise all the way to the surface. Other parts of the country are not so fortunate.

Geothermal opportunities are clustered in certain spots in other parts of the world. Volcanic activity coincides with a lot of geothermal opportunities, since the heat that fuels volcanoes can also fuel geothermal power plants. Iceland, for example, is rich in geothermal resources, since it is perched around the mid-ocean ridge in the Atlantic, the site of much volcanic activity. About 90 percent of homes in Iceland are heated with geothermal energy, and more than a quarter of the country's electricity comes from geothermal power stations.

The lack of geothermal opportunities in many parts of the world, as well as the lack of technology to take advantage of the opportunities that may exist, results in an underuse of this resource. Geothermal energy accounts for less than 1 percent of the world's energy supply, and in 2007 geothermal energy amounted to about 0.35 percent of the total supply of the United States. Increasing this percentage is an important task facing geologists and geothermal engineers.

ENERGY AND ECONOMICS

Energy is expensive. Americans spend billions of dollars on oil, some of which comes from drilling operations in American territory or just offshore, but most of which is imported from other countries. Prices

Krafla geothermal power station in Iceland *(William Smithey Jr./Getty)*

fluctuate, depending on demand. Oil prices also depend on political situations, especially in the Middle East, where there are vast reserves of oil but also much political instability. The limited supply of oil means even higher prices in the future and, eventually, depletion of this major energy resource when the oil runs out a century or two from now.

The costs of energy are not just monetary. Fossil fuel combustion releases large amounts of pollution, causing smog, acid rain—rainfall that is so acid it kills trees and plants—and an increase in respiratory and other ailments in humans. Carbon dioxide and other by-products of fossil fuel combustion may be contributing significantly to global climate change. Energy use rises along with the world's population, which now stands at more than 6 billion people, but Earth is not getting any bigger, and the environment can withstand only so much.

Developing alternative energy resources is therefore critical. Using geothermal energy would allow the United States to escape most of the volatility of the Middle East, at least in terms of oil supply and prices, and would also provide a much cleaner resource that poses far less threat to the environment. Undesirable emissions from geothermal power stations are extremely low compared to fossil fuel power generation; geothermal power stations emit only a small percent of the carbon dioxide and the chemicals released by fossil fuel power stations.

But consumer economics also plays an important role. If alternative energy sources are too expensive, many people will not buy them and in some cases cannot afford to do so. This is one of the problems holding back the progress of many green—environmentally friendly—technologies, such as zero-emission automobiles. These vehicles are much cleaner but also much more expensive than gasoline-powered vehicles, so many car buyers choose the latter. Geothermal power production is green, but it is also on average about twice as expensive as electric generators operating with fossil fuel and is less efficient—geothermal power stations extract less energy than fossil fuel power stations.

Other alternative energy technologies occasionally crowd geothermal technology out of the picture. In an Associated Press article in *Deseret News* of October 7, 2008, the Chevron executive Barry S. Andrews said, "While geothermal has gotten more attention recently, it often seems to take a back seat to solar and wind."

But the possibility of extracting a lot more energy from Earth's abundant heat is too great an opportunity to ignore. Writing in *Science* on November 30, 2007, the geologists B. Mack Kennedy of Lawrence Berkeley National Laboratory and Matthijs C. van Soest of Arizona State University noted, "It has been estimated that, within the United States (excluding Hawaii and Alaska), there are ~9 × 10^{16} kilowatt-hours (kWh) of accessible geothermal energy. This is a sizable resource compared to the total energy consumption in the United States of 3 × 10^{13} kWh annually. In order for geothermal systems to develop and mine the heat source naturally, adequate fluid sources and deep permeable pathways are a necessity."

Making geothermal energy more affordable and efficient while at the same time maintaining its environmental friendliness is a worthwhile goal. In recognition of this goal, the DOE has established the Geothermal Technologies Program, which works with the geothermal industry to lower costs and develop innovative technology. The pursuit of these objectives can take a variety of different approaches. One approach, adopted by many geothermal researchers, is to start digging.

GEOTHERMAL DRILLING

The Geysers offer a cheap source of geothermal energy that is competitive with fossil fuel power stations, but this is because the steam is easily reachable. Geothermal operations at The Geysers and Iceland are cheap and convenient because steam or hot water rises all the way to the surface, or comes close to doing so. But geothermal energy lies waiting underneath the surface at some depth everywhere on Earth. The key to exploiting this resource is to get at it cheaply and efficiently.

Oil companies obtain most of their product by drilling into the ground, either on dry land or under the ocean, and researchers looking for new geothermal resources have been doing the same. In the 1970s, DOE sponsored a series of projects in the Jemez Mountains of New Mexico, the site of hot spot volcanism and the Valles Caldera, a dormant volcano. Researchers from Los Alamos National Laboratory in New Mexico conducted tests and established a facility at Fenton Hill, New Mexico, about 37 miles (60 km) west of Los Alamos and close to the Valles Caldera. The temperature increase with depth in this area is about 186°F/mile (64°C/km), which is an extremely high rate compared to many other areas of the world.

A central theme of these tests was the concept of "heat mining"— drilling to reach Earth's hot interior. Choosing an area that has been associated with hot spot volcanic activity in the past is a wise choice, since hot rocks are probably not far from the surface. The Los Alamos researchers drilled a hole through granite rocks and reached a depth of 9,500 feet (2,900 m) in 1974, where the temperature was 387°F (197°C). In 1980, researchers drilled a deeper well, attaining a depth of 14,400 feet (4,390 m), at which the temperature rose to 621°F (327°C).

Had a steam or water pocket been available, the researchers could have built a geothermal station using the steam or water as a reservoir. But the well was dry, as expected—hot, but dry. What the researchers wanted to test was the feasibility of pumping water into and out of the hole; cool water would fall down a pipe into the bottom of the hole, where geothermal energy would heat it or boil it. Pumps would lift the hot water or steam to the surface. In this way, heat of Earth's crust could be mined as a resource.

Although the researchers performed a number of successful tests at this site in the 1980s and early 1990s, DOE cut funding to the project in 1995. The research proved that the idea was feasible from an engineering standpoint, but not necessarily an economic one. In order to offer a viable economic alternative to fossil fuel consumption, geothermal energy must keep down costs of producing the energy as well as the initial cost of establishing the facilities. Mining heat by pumping water through a well can be successful, but may or may not be the most efficient solution. There is more on this strategy in Enhanced Geothermal Systems on page 119.

Researchers have also considered drilling farther into Earth or drilling in places where water may already be present. For example, the Iceland Deep Drilling Project, conducted by a consortium of three energy companies—Hitaveita Sudurnesja, Landsvirkjun, and Orkuveita Reykjavíkur—and the National Energy Authority of Iceland, plans to drill holes up to 16,400 feet (5,000 m) to reach the hot fluid on the margins of the mid-ocean ridge. In the first test well, drilled at Reykjanes in 2004–2005, researchers went to 10,110 feet (3,082 m). The problem is that there is more heat in this region than drillers are usually equipped to handle, as temperatures may reach up to 1,110°F (600°C) at this depth.

Iceland is already blessed with significant geothermal resources, but the Deep Drilling Project, if successful, would extend their range and magnitude. Deeper wells and hotter temperatures would increase the amount of power generated from each station and the reliability and lifetime of these stations. But at these temperatures, which approach the melting point of substances such as aluminum and magnesium, geothermal engineers must learn how to protect sensitive equipment before applications can be developed.

Knowing where to drill is another important issue when considering geothermal energy. Iceland, which is located along the boundary of the separating North American and Eurasian plates, is ideally situated to take advantage of the magma and hot fluids welling up through the cracks. At other locations, geothermal researchers and developers may have little idea where to start drilling. Geothermal energy is present everywhere under the surface—Earth's interior is hot, and all a driller has to do in order to hit it is aim downward—yet the necessary depth and the presence or absence of reservoirs are often difficult to determine in advance. Drilling is expensive—a 10,000-foot (3,050-m) hole can cost a few million dollars. A geothermal developer who drills in the wrong location has wasted a lot of money.

Researchers have recently been making progress in identifying promising regions. Chapter 3 discussed helium isotopes in relation to the depth of magma, where high ratios of helium-3/helium-4 tend to be found at hot spots such as the Hawaiian Islands. B. Mack Kennedy and Matthijs C. van Soest have studied helium isotope ratios in springs, wells, and vents across a broad area covering western North America. Although high ratios are associated with volcanic regions, Kennedy and van Soest discovered high ratios in a variety of other locations. The researchers conclude that these isotopes come from mantle fluids seeping up from Earth's depths.

These locations included parts of Utah, Nevada, California, Oregon, and Idaho. The ratios rose from east to west and correspond with faults or cracks in thin layers of crust, along with ductile—bendable—crust below. This region of the country contains the Rocky Mountains, generally extending north to south, along with broad basins formed by geological activity occurring over millions of years. Volcanic activity, earthquakes, erosion, and other factors have pulled apart huge blocks of the thin crust, creating seams. The cracks extend down, but stop their descent at the more bendable crust below and start to spread sideways. These openings may allow deep fluids to percolate upward.

If the researchers' hypothesis is correct, pinpointing the best areas to drill may be as simple as looking for high helium-3/helium-4 ratios in water coming from springs or in shallow wells. Drilling an expensive series of test holes would not be necessary. Another advantage is that these sites would be excellent candidates for geothermal development, since the underground passageways would probably hold a considerable reservoir with which to work. Deep drilling may still be necessary to reach the heat and the reservoirs, but at least the developers would be confident that they are drilling in the right spot. This research, which was sponsored by DOE's Office of Geothermal Technologies, was published in a paper, "Flow of Mantle Fluids Through the Ductile Lower Crust: Helium Isotope Trends," in a 2007 issue of *Science*.

Picking the best spots to establish geothermal facilities is important to cut costs and increase efficiency. But some applications need not reach so far underground. Water reservoirs and high temperatures benefit geothermal power stations, but homeowners who want to take advantage of Earth—without leaving it worse for wear—can rely on a lot less. As mentioned earlier, geothermal energy provides heating for the majority of homes in Iceland. People in the United States and other countries can also keep their houses comfortable with a little help from Earth's interior. The technology necessary is called a geothermal heat pump.

GEOTHERMAL HEAT PUMPS

Thermodynamics—the flow of heat—does not tend to work to a homeowner's advantage. Heat spontaneously flows from hot objects to cool ones, or, in other words, heat goes from hot objects to cool ones unless something prevents or otherwise influences this natural course of action. In the summer, heat seeps into a cool house, raising the temperature to an uncomfortable level unless the owner turns on the air conditioner. In the winter, heat from a roaring fire or the heater escapes, forcing the owner to bundle up or stoke the fire. Maintaining a comfortable temperature in the house requires an expenditure of energy, in the form of electricity to run the air conditioner or fuel for the heater. Utility bills in the winter and summer can be expensive.

An air conditioner would seem to defy the laws of thermodynamics since it cools a house by transferring heat to the outside, which in the summer is warmer than the house. Air conditioners can function because they use fluids circulating in the system that alternately expand, capturing heat from the house, and then condense in pipes outside, which transfers the heat. The expenditure of energy comes from the need to include a compressor, which is usually an electric motor that compresses the fluid before it reaches the condenser. Compressing the fluid increases its temperature; by warming the fluid so that its temperature is higher than the outside air, heat naturally flows out of the condenser. Once the heat flows out, the fluid cools, and returns inside the house to pick up more heat for the next cycle. The laws of thermodynamics are satisfied, but the homeowner is left with a huge electric bill.

Much less energy would be needed to operate the system if a large object was available to absorb heat during the summer and supply heat in the winter. This object must have a steady temperature in the comfort zone—not too cold and not too hot—throughout the whole year, so that heat would flow out of the house in the summer and flow into the house during the winter. The object must also be large enough so that the heat flowing into or out of it would not be enough to change its temperature. Such an object actually exists—Earth's crust, at a shallow depth.

The temperature of Earth's crust rises rapidly with depth, but just beneath the surface there is a stable zone that stays at or near the same temperature all year. Cave visitors are aware of this, since the temperature inside large, deep caves such as Carlsbad Caverns in New Mexico and Mammoth Cave in Kentucky stays around 55°F (13°C) even in the sizzling summer. Geothermal heat pumps work by using Earth to absorb excess heat in the summer—in this capacity, Earth is sometimes called a heat sink—and supplying heat in the winter.

Any homeowner can take advantage of this technology—geothermal heat pumps do not require the presence of underground water reservoirs or steam—and thousands of geothermal heat pumps have been installed in the United States. (Some kinds of air-conditioning systems are known as heat pumps, but these systems should not be confused with geothermal heat pumps, as they do not draw on geothermal energy.) For a house with a large yard, geothermal systems can employ a lot of horizontal pipe, which only needs a depth of about five feet (1.5 m). Homeowners with small yards may opt for a vertical system, in which the pipes reach 100–300 feet (30.5–91.4 m). Systems for larger buildings, such as schools and office complexes, must be more extensive, but in any case the underground pipes are covered and do not interfere with the use of the surface area. The pipes are usually made of a durable plastic that is an effective thermal conductor, permitting the exchange of heat between Earth and the system fluid. A geothermal heat pump operates similarly to an air conditioner, transferring heat from inside the building to the outside—in this case, underground—during the summer. The operation is the same in the winter, but works in the reverse direction.

Geothermal heat pumps are about twice as expensive to buy and install as conventional heating and cooling systems, but about two-thirds of the energy comes from Earth and is green energy. Another advantage is a reduction in the energy bill. The amount of the reduction varies, depending on the cost of energy in the area; in many cases, utility bills are cut in half, and the system also supplies hot water.

Although many homeowners and businesses have taken advantage of geothermal heat pumps, these systems are not very common. Once again, the higher initial cost is a barrier. Perhaps the newness and unfamiliarity of geothermal energy also contributes to consumer hesitation.

But as the population continues to grow and make increasing demands on Earth's natural resources, alternative sources of energy must take the place of fossil fuels. Geothermal energy, whether in the form of heat pumps and other small systems for buildings, or in the form of geothermal power stations that serve cities, is an option that geologists wish to explore further.

Some of the most important issues concerning alternative energy supplies, including geothermal energy, are their potential and capacity. The best energy sources to pursue are those that may have the capacity to supply a large amount of energy, if this potential can be realized economically.

GEOTHERMAL POTENTIAL AND CAPACITY

The Mid-Atlantic Ridge and the Pacific's Ring of Fire offer many geothermal energy opportunities, as do hot spots such as Hawaii. Geothermal energy's potential and capacity depend on how easily and efficiently these and other sources can be harnessed.

Some of the problems with geothermal power stations are due to the same reason why hot springs are often popular settings for health spas—the presence of minerals in the water. Silica, the sandy material that is used for glassmaking and other applications, is a common constituent of geothermal reservoirs. Hot water carries a substantial amount of silica, dissolved from the surrounding rock. As geothermal systems extract energy from the water, its temperature drops, and silica precipitates out of the solution as a solid. These glassy solids get deposited in the pipes and heat transfer systems, reducing the flow of water and sometimes even clogging the pipes completely. Technicians must periodically remove these deposits, or the energy conversion process will be impaired and become inefficient. Maintenance of the pipes and other, more delicate parts of the system adds greatly to the cost of geothermal power stations.

Geologists are not sure how the silica precipitates, making it hard to control this process. But recently an Ohio University researcher Dina Lopez and her colleagues studied some geothermal stations in El Salvador in Central America. Located on the Ring of Fire, El Salvador is a prime spot for geothermal energy. Lopez and her colleagues examined the silica problem and developed a model, which they presented at the 2007 meeting of the Geothermal Resources Council, an international association of geothermal scientists and developers. The model uses advanced principles of geochemistry, but also incorporates experiments with silica formation—how fast it builds up given the conditions inside the geothermal stations. With this model as a guide, technicians can gauge the rate and impact of this problem in their geothermal systems. Combined with further studies of geothermal power stations, the model will help builders and operators design more efficient systems by showing when and where silica is likely to form.

In addition to efficiency, concerns about geothermal energy include possible depletion of the resource, similar to what is happening with fossil fuels. The supply of fossil fuels is limited and will be exhausted in the not too distant future at the present rate of consumption. To find alternatives, researchers are emphasizing energy sources that are not only green but also renewable or, in other words, resources that do not come in a limited, exhaustible supply. Solar energy is renewable, for example, because the Sun will continue to shine for billions of years.

Is geothermal energy renewable? A geologist could argue that geothermal energy is not renewable because Earth holds only a certain amount of heat. As heat flows out of an object, its temperature drops. Excessive use of geothermal energy could possibly cool the planet's accessible depths to an unusable temperature.

But from a practical standpoint, renewability comes down to the rate of use versus the rate of generation. Radioactive decay within Earth will continue to supply heat, which will fail only in the remote future, when all the radioactive isotopes decay into stable, nonradioactive isotopes. And since the planet is so large, Earth already contains an immense quantity of energy. Geothermal energy is renewable if the demands are not too large.

Depletion at individual power stations is a concern, however. For geothermal power stations relying on hot-water reservoirs, strenuous use could lower the temperature too quickly for it to be recharged by heat flowing from the surrounding areas. Heat does not conduct very

Interdisciplinary Science—Many Specialties, One Goal

Most scientists spend part of their time directly on science—making observations in the field, doing experiments, or developing theories—and the rest of their time fulfilling other obligations, such as writing reports in their area of expertise, evaluating the reports of other scientists that have been submitted to scientific journals (so that the journal editors can decide whether the report is worthy of publishing), and participating in conferences. Scientists contribute their expertise in the writing and evaluation of reports to increase knowledge in their own disciplines, but scientists with different specialties often attend the same conferences. Such meetings provide opportunities to exchange ideas and to learn different points of view.

An interdisciplinary panel, such as that organized by MIT to study geothermal energy, is a means of exchanging information and pooling expertise in the pursuit of a common goal. No one can possibly have a deep knowledge of all fields of science and engineering; more than half a million scientific papers are published each year, and keeping up to date on just a single branch or discipline is demanding enough.

quickly through rock, so in the absence of significant convection currents, it may take decades for a cooled reservoir to reheat.

Efficiency, reservoir depletion, drilling costs, and the lack of accessible reservoirs at certain geothermal sites are factors that limit geothermal energy's capacity and potential. But just how limiting these factors really are depends on what kind of technology scientists and engineers may be able to develop to surmount them. An international panel of scientists and engineers from a variety of fields, organized by the Massachusetts Institute of Technology (MIT), recently addressed these issues and issued a report, *The Future of Geothermal Energy,* in 2006.

Heading the MIT panel was Jefferson W. Tester, an MIT professor of chemical engineering (the study of chemical reactions and conversions that produce industrially useful substances). The panel also included chemists, geophysicists (geologists who specialize in applying the principles of physics to the study of Earth), engineers who specialize in petroleum products, geothermal experts, economists who specialize in the study of energy production and management, and experts in the conversion of energy into electrical power. All these specialties were important in analyzing the problem of developing geothermal energy into the most useful products in the most efficient way.

Organizing the MIT panel, bringing the panel members together, and providing the essential data and materials to study costs money. The Office of the Geothermal Technology Program, established by DOE, donated these funds. Panel members met and reviewed past and current research projects from the United States, Europe, Japan, and Australia. The panel's findings were detailed in a 372-page report, *The Future of Geothermal Energy,* issued in 2006. This project serves as an example of the need to call on the skills of many different people in order to tackle a complicated scientific or technical problem.

The report was the product of a team of 18 scientists, engineers, geothermal specialists, and drilling experts. As described in the side-bar on page 116, interdisciplinary collaboration—specialists in different disciplines or fields of study working together—is often necessary to solve complex problems, because the span of such problems, as well as potential solutions, covers more subjects than any one person can learn. Topics important in the development of geothermal energy include geology, chemistry, economics, drilling technology, thermodynamics, and a variety of engineering specialties.

One of the main thrusts of the report is the need to expand or en-hance geothermal technology. Systems that take advantage of accessible reservoirs, such as those in California and Nevada, are already in op-eration, but the energy yield is not yet a significant contributor to the nation's energy resources. There is a wealth of geothermal energy that is not quite as accessible, but that if tapped would greatly increase geo-thermal use. The MIT panel focused on what scientific and technologi-cal innovations are necessary to make the development of these hard-to-reach resources economically feasible.

Research at Fenton Hill, New Mexico, in the 1970s through the middle of the 1990s featured prominently in the MIT panel's analysis. As previously described, this project successfully used geothermal ener-gy to heat water pumped through deep holes, instead of relying on wa-ter reservoirs below ground or steam rising to the surface. This pumped water became the carrier of heat or, in other words, the heat transfer mechanism by which scientists extracted energy from Earth's interior.

The MIT panel suggested the continuation of the Fenton Hill project and similar field experiments. Not enough people in the 1990s were interested in investing money in a geothermal power station at Fenton Hill, but the MIT panel concluded from their technical and economic analyses that such investments would pay off. In *The Future of Geothermal Energy,* the researchers wrote, "Based on growing mar-kets in the United States for clean, base-load capacity, the panel thinks that with a combined public/private investment of about $800 mil-lion to $1 billion over a 15-year period, EGS [enhanced geothermal system] technology could be deployed commercially on a timescale that would produce more than 100,000 MWe [megawatts electric, a quantity of power] or [equivalently] 100 GWe [gigawatts electric] of new capacity by 2050." This amount of power is about 10 percent of today's generating capacity, so it is a considerable quantity. Although $1 billion is a lot of money, it is comparable to construction costs of a single major power station.

Other researchers have voiced similar conclusions. The National Renewable Energy Laboratory (NREL), a research and development facility of the DOE, issued a report after holding a workshop in Golden, Colorado, on May 16, 2006. The report, *Geothermal—the Energy under Our Feet,* authored by Bruce D. Green and R. Gerald Nix of NREL, summarized the potential of geothermal energy by writing, "The energy content of domestic geothermal resources to a depth of 3 km (~2 mile) is estimated to be 3 million quads [1 quad = 170 million barrels of oil], equivalent to a 30,000-year supply of energy at our current rate for the United States!"

In order to obtain this energy, researchers must find a way to transport it to the surface. This job requires the design and development of enhanced, or engineered, geothermal systems.

ENHANCED GEOTHERMAL SYSTEMS

The project at Fenton Hill, New Mexico, was initially known as a hot dry rock project, since the goal was to extract subsurface energy without benefit of an existing water reservoir. But the term *enhanced geothermal systems* can encompass a wide range of technologies to enhance the amount of energy obtainable by geothermal systems. In some cases, underground water reservoirs are available but are difficult to reach or are surrounded by rocks with low permeability, which means little water can seep through them. Gaining access to this reservoir requires increasing rock permeability, perhaps by creating or widening cracks in the rocks with mechanical force or exposure to harsh acids. Higher flow rates enhance the quantity of available energy.

Hot dry rock technology introduces water into the system by pumping cool water down a drilled hole. The pump, operating at high pressure, forces water through a zone of high temperature. After channeling through this zone, the heated water may be suctioned or forced out another drilled hole, which brings the heat energy to the power station or some other geothermal conversion device.

In some cases, the optimism of the MIT panel and NREL workshop has been substantiated. For example, in February 2008, Geodynamics

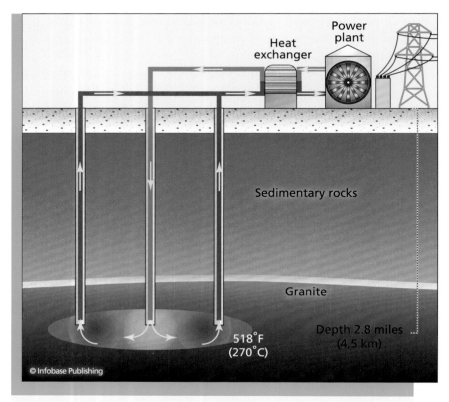

In this geothermal system, water circulates in pipes, which extend through the relatively cool sedimentary rocks to reach the hot granite rocks below. After heating, the water enters the heat exchanger and releases some of its energy, which drives the electricity generators in the power plant.

Ltd., an Australian energy company, completed drilling a 13,845-foot (4,221-m) well at Cooper Basin in South Australia. This project, called Habanero 3, is one of three wells the company has drilled. The company plans to build a large-scale power station at this spot using technology they refer to as hot fractured rock.

Cooper Basin is an excellent location for this project. Much of this area is a desert lying above a granite rock bed having a temperature of about 480°F (249°C) at a depth of less than 2.5 miles (4 km). The region therefore has a lot of heat available at a shallow depth—it is somewhat like a hot spot without the volcanic activity (see chapter 3).

As illustrated in the figure opposite, Geodynamics plans to pump cool water through the top-level sedimentary rocks and into the hot, fractured granite below. The water picks up heat as it travels through the hot rocks and is taken back up to the surface, where it releases its energy in a heat exchanger. This energy runs an electric power station. Geodynamics engineers say that a volume of rock of about 0.24 cubic miles (1 km^3) at this temperature contains the same energy as 40 million barrels of oil. The company believes this geothermal energy can produce electricity as cheaply as fossil fuel power stations, with the benefit of greatly reduced emissions.

Engineered geothermal systems are also in the works in Europe and in the United States. A large number of organizations and companies announced a partnership in February 2008 with plans to test a system at Reno, Nevada, at a well near an existing geothermal power station owned by Ormat Technologies, Inc. In addition to Ormat, which is based in Nevada, the partners include DOE, the engineering firm Geothermex, Inc., of California, as well as researchers from the United States Geological Survey (USGS), Idaho National Laboratory, Sandia National Laboratory in New Mexico, Lawrence Berkeley National Laboratory in California, and others.

The problem with the well is not its location, as evidenced by the nearby geothermal power station. This well's problem is that in its current state it does not produce enough hot water. To enhance the supply, researchers plan to test methods of increasing the permeability of the underground rocks. The additional flow of water may create a viable geothermal system out of a well that is not now economically feasible to operate. With the addition of the enhanced system, Ormat believes the site can produce about five times the current capacity of the nearby (unenhanced) geothermal power station.

Other companies are also investing in geothermal energy. Google, a computer technology company that owns one of the most widely used search engines on the Web, announced on August 19, 2008, that it had invested $10.25 million in enhanced geothermal systems. Dan Reicher, the director of climate and energy initiatives for the company's philanthropic division, said in a press release, "EGS could be the 'killer app' of the energy world. It has the potential to deliver vast quantities of power 24/7 and be captured nearly anywhere on the planet. And it would be a perfect complement to intermittent sources like solar and wind."

CONCLUSION

As the world's supply of fossil fuel comes to an end, alternative energy sources must be found. To the extent that these sources are renewable, people will not face expensive and disruptive energy crises in the future. Energy production and consumption that emit little pollution or otherwise entail minimal damage to the environment or climate are also a high priority. Geothermal energy offers one potentially large resource that has yet to be fully developed. Increasing the world's underused geothermal energy capacity, either by taking advantage of accessible steam or hot-water springs or by engineering more sophisticated systems, would greatly contribute to the solution of energy shortages, rising costs, and environmental concerns.

But developing enhanced geothermal systems will not necessarily be easy. A program in Switzerland recently encountered serious trouble. This project, launched by the Swiss company Geopower Basel, aims to extract geothermal energy 3.1 miles (5 km) below the city of Basel by injecting water at high pressure to absorb the heat. The concept is the same as the Australian project at Cooper Basin mentioned earlier. But on December 8, 2006, as the company was in the process of testing their system by injecting water deep below ground, Basel was hit with a minor earthquake measuring 3.4 on the Richter scale. (The Richter scale, as described in chapters 1 and 6, is the old but still sometimes used method of quantifying the magnitude of earthquake waves.) Although a 3.4 earthquake is small, some buildings in Basel sustained damage. Testing stopped at once. Even so, in the weeks that followed, several more tremors struck the city.

Several hundred earthquakes shake Switzerland every year, most of them quite minor. Basel has had more than its share of these tremors. For example, an earthquake estimated to be 6.5 in magnitude devastated the city in 1356. Engineers associated with the geothermal project voiced some concerns about the seismic fault and earthquake activity in the area, but the size and number of tremors apparently triggered during the system tests were not anticipated.

Although the project has not been abandoned, company and city officials postponed any further work until geologists and engineers finish studying the exact cause of the tremors. The results of these studies will be used in a thorough analysis of the risks of the project, which is unlikely to be resumed for several more years. If the plan is ultimately successful, the geothermal energy would provide heat or electricity to about 10,000 homes. To Basel, with its population of about 160,000 people, this energy contribution would be significant.

Enhanced geothermal systems can extend the economic use of geothermal energy to regions that would otherwise have difficulty accessing this energy, but triggering earthquakes or adversely affecting ground stability is not acceptable. Some areas of the world are more susceptible to this risk than others. In Basel, geologists may determine that the danger is too great, especially since the operation would be conducted under a populous city. The deserts of Australia and Nevada seem to be much less susceptible, and such problems are not likely to arise.

As with all the proposed alternative energy sources, the costs and risks will continue to be determining factors in where and to what extent people can effectively use geothermal energy. Other alternative energy projects, such as those involving wind, wave, solar, nuclear, and hydroelectric power, are important but may drain too much funding away from geothermal research and development, limiting or disabling future projects. The needed replacements for fossil fuels have yet to be established, so the best course of action is probably to pursue as many options as possible. Geothermal energy is an outgrowth and a frontier of Earth science whose potential is as big as the planet.

CHRONOLOGY

79 C.E.	Romans have by this time, and perhaps much earlier, developed elaborate plumbing systems in the city of Pompeii and elsewhere to use geothermal energy to heat their homes and baths.
1847	Surveyors led by John C. Fremont (1813–90) discover The Geysers, an area of California rich with steam and hot springs rising from the surface.

1890s	Boise, Idaho, develops the first geothermal district heating system, in which water from hot springs is piped into some of the city's buildings.
1904	Working in Larderello in central Italy, the Italian chemist and inventor Piero Ginori Conti (1865–1939) builds the first electric generator running from geothermal power.
1930	Iceland begins the widespread use of geothermal energy to heat buildings.
1960	Pacific Gas and Electric Company opens the first geothermal power station in the United States at The Geysers in California.
1970s	The U.S. DOE initiates a series of geothermal research projects in the Jemez Mountains of New Mexico and establishes a testing facility at Fenton Hill, New Mexico.
1980s	Geothermal power stations appear in Hawaii, Nevada, and Utah.
2006	An interdisciplinary panel of scientists and engineers, organized by MIT, issues a report, *The Future of Geothermal Energy,* which recommends investment in enhanced geothermal systems.
	Bruce D. Green and R. Gerald Nix of NREL, issue a report proposing research on geothermal systems, based on the findings of a workshop in Golden, Colorado, on May 16, 2006.
2008	An Australian company Geodynamics, Ltd., completes drilling a well at Cooper Basin in South Australia and begins development of a geothermal power station using hot fractured rock technology.

FURTHER RESOURCES
Print and Internet

Boyle, Godfrey, ed. *Renewable Energy,* 2nd ed. Oxford: Oxford University Press, 2004. After an initial chapter introducing renewable energy, the book describes solar energy and technologies, energy from biological sources, hydroelectricity, tidal energy, wind energy, wave energy, and geothermal energy.

California Energy Commission. "Geothermal Energy." Available online. URL: http://www.energyquest.ca.gov/story/chapter11.html. Accessed May 4, 2009. California generates more electricity from geothermal energy than any other state. This Web page, which is one of the chapters in the California Energy Commission publication "The Energy Story," contains a concise description of geothermal energy.

Deseret News. "Geothermal Energy Development Gathers Steam." October 7, 2008. Available online. URL: http://www.publicutilities.utah.gov/news/geothermalenergydevelopmentgatherssteam.pdf. Accessed May 4, 2009. This short article sounds an optimistic note on geothermal energy.

Geothermal Education Office. "Geothermal Energy." Available online. URL: http://geothermal.marin.org/. Accessed May 4, 2009. The Geothermal Education Office, a nonprofit organization in California to promote awareness of geothermal resources, offers a lot of information on the Web. Topics include geothermal facts and a glossary of terms, a slide show, general information about energy resources, and maps showing the worldwide use and application of geothermal energy.

Gibilisco, Stan. *Alternative Energy Demystified.* New York: McGraw-Hill, 2007. In addition to explaining the concept of energy and energy transformations, this book covers a broad range of energy technologies. The author's classification of geothermal power as an "exotic" means of electricity generation is perhaps not the best word choice, but readers can compare and contrast a variety of fossil fuel and alternative energy sources.

Google. "Google.org Invests More Than $10 Million in Breakthrough Geothermal Energy Technology." News release, August 19, 2008.

Available online. URL: http://www.google.com/intl/en/press/press-rel/20080819_egs.html. Accessed May 4, 2009. Google announces a significant investment in geothermal energy.

Green, Bruce D., and R. Gerald Nix. "Geothermal—the Energy under Our Feet." Available online. URL: http://www1.eere.energy.gov/geothermal/pdfs/40665.pdf. Accessed May 4, 2009. This technical report is based on a workshop held at Golden, Colorado, on May 16, 2006, sponsored by NREL. Web users with slow download rates should be warned that the size of this file is about 2.8 megabytes.

Kennedy, B. Mack, and Matthijs C. van Soest. "Flow of Mantle Fluids through the Ductile Lower Crust: Helium Isotope Trends." *Science* 318 (November 30, 2007): 1,433–1,436. The researchers report on their measurements of helium isotopes ratios in springs, wells, and vents in a broad area covering western North America.

Massachusetts Institute of Technology. "The Future of Geothermal Energy." Available online. URL: http://geothermal.inel.gov/publications/future_of_geothermal_energy.pdf. Accessed May 4, 2009. MIT's geothermal energy report that was discussed in this chapter is available on the Web. But Web users with slow download rates should be warned that the size of this file is about 14 megabytes.

ScienceDaily. "A Step toward Inexpensive Geothermal Energy." News release, March 15, 2007. Available online. URL: http://www.sciencedaily.com/releases/2007/03/070313110634.htm. Accessed May 4, 2009. Dina Lopez, a researcher at Ohio University, reports her findings on geothermal stations in El Salvador and ways to improve efficiency.

Smil, Vaclav. *Energies.* Cambridge, Mass.: MIT Press, 1999. This book offers a general view of energy, describing its many forms, its scientific properties, and the manner in which it is used.

Swissinfo.ch. "Man-Made Tremor Shakes Basel." December 9, 2006. Available online. URL: http://www.swissinfo.org/eng/news/science_technology/Man_made_tremor_shakes_Basel.html?siteSect=514&sid=7334248&cKey=116583965 8000&ty=st. Accessed May 4, 2009. Geothermal drilling is suspected to play a role in the minor earthquake that struck Basel, Switzerland, on December 8, 2006.

Union of Concerned Scientists. "How Geothermal Energy Works." Available online. URL: http://www.ucsusa.org/clean_energy/technology_

and_impacts/energy_technologies/how-geothermal-energy-works.html. Accessed May 4, 2009. This introduction to geothermal energy contains three sections: the geothermal resource, how geothermal energy is captured, and the future of geothermal energy.

United States Department of Energy. "Geothermal Technologies Program." Available online. URL: http://www1.eere.energy.gov/geothermal/. Accessed May 4, 2009. The Web resource describing DOE's program to enhance geothermal applications includes basic information, maps, photographs, data on current exploration and research, and a brief history of the subject.

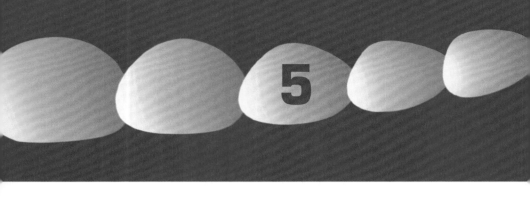

WATER MANAGEMENT— CONSERVING AN ESSENTIAL RESOURCE

Water management is not a new problem. Ancient Rome was a bustling, crowded city, housing more than a million people. A city this size is large even by modern standards, and for the ancient world, a city with a million inhabitants presented numerous sanitary and engineering challenges. One of the most important concerns was freshwater. The Tiber, a river that flows through Rome, supplied plenty of water during the city's early years, but as the population grew, the Tiber became inadequate as well as polluted. Wells and rainwater were also insufficient, so the Romans had to carry water into the city from springs and other distant sources. To accomplish this task efficiently, Roman engineers began building an ingenious system of long channels or conduits—aqueducts—in 312 B.C.E.

The idea behind an aqueduct is simple. An aqueduct is made of stone or concrete and has a gentle slope, so that water in the channel flows downhill from the source—a spring on a hill, perhaps—all the way to the city. But engineering these aqueducts required considerable skill, for the channel must maintain its slope as it winds its way along the countryside. The Romans used natural slopes such as a hillside wherever possible and built arched bridges when necessary. A total of 11 aqueducts eventually served the city, the longest having a length of about 60 miles (37 km).

Freshwater continues to be a concern for cities today. Humans need roughly a half gallon (1.9 L) of water a day to survive, and most people use a lot more than that for washing, sanitation, and other applications. But according to the World Health Organization (WHO), the number of people in 2002 who did not have access to safe drinking water exceeded 1 billion—about 17 percent of the world's population—and thousands of people die every day from waterborne illnesses such as bacterial infections. A growing population will put even more pressure on scarce water resources. A United Nations (UN) report, "Water for People, Water for Life," issued in

Crop irrigation is often essential for adequate yields. *(Don Bayley/iStockphoto)*

2003, predicts about half of the world's population will experience significant water shortages or other major problems in the next 50 years: "Critical challenges lie ahead in coping with progressive water shortages and water pollution. By the middle of this century, at worst 7 billion people in sixty countries will be water-scarce, at best 2 billion people in forty-eight countries."

Although Americans generally have access to plenty of *potable* water, shortages caused by a drought can occur in various cities and states. In the 1930s, a severe drought struck the lower Midwest, parching the soil and resulting in the dust bowl, as the wind blew clouds of stifling dust throughout the region. Because freshwater is essential to life, people in all countries should be aware of issues concerning shortages of clean, drinkable water. One of the most important goals of Earth science is to understand the dynamics and distribution of this life-giving resource and to find ways of alleviating present and future shortages.

As geologists gain a more complete understanding of Earth and its water distribution, locating and exploiting water resources such as underground sources will become easier and cheaper. But there is also an urgent need to understand and predict rainfall variations and climate patterns so that events such as droughts do not catch people off guard. An extra complication is that the world's climate has been changing recently, with temperatures rising about 1.3°F (0.74°C) on average in the last century, and this change will probably have a strong effect on the planet's water circulation. Studying any phenomenon on a scale as large as an entire planet is challenging because so many factors come into play. Coping with this complexity is one of the important frontiers of Earth science. This chapter discusses how researchers are learning more about the ways in which water cycles through the environment and how to use that knowledge for improved water management.

INTRODUCTION

Water is a compound made of two parts hydrogen (H) and one part oxygen (O). The chemical formula is H_2O. The bond between the hydrogen atoms and the oxygen atom is covalent, which means that bonded atoms share electrons. Oxygen's nucleus pulls on the shared electrons with a little more strength than hydrogen's nucleus, so the electrons tend to be closer to the oxygen atom than the hydrogen atoms. This unequal charge distribution means that a water molecule is not electrically neutral, so it attracts other charges. Water molecules tend to stick together because of this attraction and also have the ability to pry apart compounds such as salt, dissolving them—water is an excellent solvent.

The ability to dissolve many kinds of substances makes water a good choice for cleaning and washing purposes. For similar reasons, water is an effective medium for living organisms, dissolving and carrying nutrients and other required molecules. Water comprises about 60 percent of the weight of an average person. Life on Earth evolved in the seas, which are salty because of dissolved substances such as sodium and chloride ions (charged atoms that result when sodium chloride dissolves in water). Living organisms retain a portion of this environment within them—certain fluids in the body of humans and other organisms are chemically similar to the ocean, containing large quantities of sodium and chloride ions.

The body is constantly losing H_2O, and this water must be replaced. People must drink freshwater instead of salt water in order to maintain the delicate balance of ions and other substances in the body. (Certain marine mammals such as seals can drink salt water, but, unlike humans, they have the ability to excrete the excess ions.) Potable water must also be free of disease-causing microorganisms that can invade intestinal tracts and result in serious problems in digestion and elimination.

Although people can survive with only a half gallon (1.9 L) per day, many people typically use more than this for bathing and washing. Businesses also consume a large amount of water during cleaning and other operations. Another major use of water is crop irrigation, which is essential in drier parts of the world. About a third of the water supply in the United States gets used in farming, and this percentage is larger in states such as California and Texas.

Earth is the only planet in the solar system with an abundance of H_2O in three phases—as a gas (water vapor), liquid (water), and solid (ice). About 70 percent of the planet's surface is water, and there is also much water beneath the surface. The total volume of water on Earth is 326,000,000 cubic miles (1,360,000,000 km³). These units are large—a cubic mile (4.1 km³) is a cube having one mile (1.6 km) per side—and millions of them add up to a great deal of water, enough to form a body of water the area of the United States with a depth of 90 miles (145 km)! Oceans and seas contain about 96 percent of this water, so only about 4 percent of Earth's water is fresh, much of which is locked up in glaciers and polar ice.

The total amount of water on Earth does not change much—some molecules escape Earth's gravity, and some get broken down into their constituent elements, but new water arises from chemical reactions or is brought by comet impacts. The overall quantity can change from time to time but not by a lot. Water does move around a lot, though, and change phases.

Where did all this water come from in the first place? Geologists are not yet sure of the answer. Some of Earth's water was probably mixed in with the material that originally formed the planet. Much of this water would have been a component of aggregates and minerals; heat given off during the planet's creation melted the material and freed the water, which rose to the surface through volcanic activity. But some of Earth's water may have arrived from space. For example, the impacts of comets,

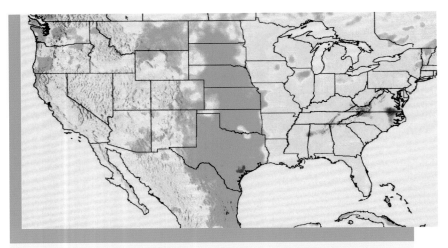

Rainfall in the United States for the first eight months of 2007—blue and green indicate a higher than average rainfall whereas yellow and orange indicate less than average. The Midwest was more wet than usual in this period. *[NASA image by Hal Pierce [SSAI/NASA GSFC]]*

which mostly consist of ice and frozen gases such as carbon dioxide, may have brought some additional water to the planet. In either case, the water is here to stay—little of it escapes or gets broken down.

But Earth's water is not distributed equally. *Hydrology* is the study of the properties, distribution, and circulation of water. (*Hydro* comes from a Greek term meaning "water.") Some regions are dry, such as the Sahara in Africa, parts of which receive an average of only about three inches (7.5 cm) of rainfall a year. Other regions are wet, such as rain forests around the Amazon River in South America, parts of which average about 200 inches (500 cm) of rainfall a year. Rainfall can also vary widely in the same region at different times of the year—some regions, such as parts of Asia, experience a dry season and an extremely rainy one. And rainfall usually varies considerably in the same region from year to year.

As a result of variations in rainfall, people throughout history have made concerted efforts to stabilize their water supply. Cities have often been founded near lakes or rivers, which have also provided useful means of transportation. Springs can also be an important source of water. When none of these sources suffice or they become too polluted, citizens must turn to other strategies, as the Romans did when they built aqueducts.

Traditional means of increasing the water supply include piping water from the surrounding area, digging wells, and blocking rivers with dams to create lakes. While effective, these methods have become increasingly costly because expanding populations require ever larger sources, often at great distances from cities. There are also environmental concerns, such as when dams impede the flow of rivers and flood important wildlife habitats.

THE WATER CYCLE

A fundamental concept of hydrology is the water cycle. Although the total amount of water on the planet is relatively constant, water does not stay in one place, or in one form, but cycles between various reservoirs—sites at which water is stored. (The largest water reservoirs on Earth are the oceans.) Water moves from reservoir to reservoir; for example, molecules of water evaporate, leaving the liquid phase to become water vapor, and then condense, the opposite process by which gaseous molecules become a liquid. Water vapor, being a gas, is not visible, but condensation often occurs around small particles suspended in the atmosphere, forming clouds that are visible because of the water droplets or ice crystals, if the water droplets are cold enough to freeze. This condensed water eventually falls as *precipitation*—rain, snow, hail, or sleet. In this manner, water molecules cycle from surface water to atmospheric vapor to precipitation, at a rate that varies widely from molecule to molecule. Some molecules may complete a cycle in a day or two, others may take years.

The driving force of the water cycle is the Sun. Sunlight heats the water surface, giving H_2O molecules sufficient energy to escape into the air. (The water molecules escape, but most of the other substances, such as salt in the ocean, stays behind.) This water vapor joins the atmosphere, eventually to condense into water droplets of around 0.0004 inches (0.001 cm) in diameter, forming clouds. Clouds are white because these droplets reflect all of the wavelengths of sunlight; none of the colors predominate, so the reflection is white. These tiny droplets can remain suspended in the air, carried by winds, until they grow much larger and fall to the ground as precipitation. Clouds become dark when they thicken and block the sunlight—these are often storm clouds that cause a lot of precipitation.

Many people would like to control rainfall—for example, farmers whose crops need watering. Although some engineers try to seed clouds by dispersing tiny particles on which water can condense, this can only

work if there is water vapor in the air. The amount of water vapor in the air is called *relative humidity,* given as a percentage of the maximum amount of water vapor that air can hold. This maximum depends on the air's temperature—warm air can hold more than cold air. If the temperature is 70°F (21°C), for example, and the relative humidity is 60 percent, then the air is holding six-tenths the maximum at this temperature. Highly humid conditions feel muggy. During the night and early morning, as temperatures cool, excess water vapor in the air condenses, forming dew.

Much of the water cycle takes place as water evaporates from the warm, tropical regions of the oceans. This warm, humid air moves inland, cooling as it encounters colder territory, particularly if it has to rise over mountains. Cool air can hold less water, resulting in precipitation. The process is similar to a sponge—warm air absorbs water, and cooler temperatures wring it out.

Precipitation falling on the ground meets one of two different fates. Most of this precipitation is water or if ice or snow eventually melts and becomes water. Some of the rain or snowmelt flows along the ground or in the gutters of cities, eventually reaching a stream or river. Rivers flow into oceans, delivering the water back to where it started for another cycle. The rainwater that flows over the surface is known as runoff. (In addition to contributing to the water cycle, running water also plays an important role in shaping the surface of the planet through erosion and weathering.) Water that accumulates in ponds and lakes stays around for a while, although molecules on the surface of these small bodies of water evaporate and continue to participate in the cycle. The figure opposite illustrates the water cycle.

The other possibility is that the water will be absorbed into the ground in a process known as infiltration. In this process, water molecules seep between soil particles or through tiny pores or cracks in rocks. About 1 percent of Earth's water supply exists below the ground, constituting a quarter of the planet's freshwater; only a tiny fraction of freshwater remains on the surface in the form of lakes and rivers.

(opposite page) Water molecules evaporate from the oceans and other bodies of water and then fall as precipitation. The cycle starts anew as water drains into lakes, rivers, and seas.

Some cities draw at least a portion of their freshwater from nearby rivers or lakes. The faucets of people in Philadelphia, Pennsylvania, for instance, deliver filtered and treated water taken from the Schuylkill

Aquifers—Underground Water Sources

Water can flow underground, as it does aboveground, but not nearly as fast since it must seep through soil or rocks except in rare cases when a tunnel or cave is present. Many underground regions that have a lot of pores or crevices allow at least some flow, so groundwater can seep into a well to replace some of the water that has been pumped out. But if water is pumped out faster than it seeps in, the water level drops. This drop may be enhanced by a long dry spell, decreasing the amount of available groundwater. Even if the water table is close to the surface in the region of the well, not enough water can seep in fast enough to replace the large quantity that has been pumped out. The well depresses the water table in the immediate vicinity, as shown in the following figure, and can go dry.

Excessive pumping can also cause the surface around the well to collapse. The reason for this collapse is the loss of water in the pores and crevices, which creates empty spaces under the ground. As sediment falls into these spaces, the volume decreases—the material is more compact—and the surface becomes lower. If the water loss opens up a significant amount of space, the material may not be able to support the weight of the soil and rocks above. The result is a sinkhole or fissure.

When too much groundwater is pumped from a larger area, such as around a large city, the land can subside—shift downward. Subsidence is a big problem in cities that pump groundwater to satisfy a substantial portion of their water needs. In Mexico City, land subsidence caused by groundwater withdrawal has resulted in warped roads and damaged buildings and has induced a slight lean in the beautiful Metropolitan Cathedral.

River and the Delaware River. Other communities rely on ground-water, which is pumped from wells. For example, the majority of the water supply in Mexico City, Mexico, comes from groundwater. An

When users pump a well faster than nearby sources can replenish it, the water level drops around the well, forming a cone in which the water table is depressed.

underground area that has enough water to supply a well is known as an *aquifer.*

There is a lot of water underground, but the big question is how far well drillers need to go before they strike pay dirt (which is wet, in this case). Gravity pulls down any water that infiltrates the ground. Soil or rock just beneath the surface is usually dry (except on a rainy day) and is known as the unsaturated zone or *vadose zone.* At some point below this zone, water fills every pore and crevice, saturating the ground with water—this is known as the *saturated zone.* (But this zone only reaches a short distance below the surface. The pressure of deeper depths is too great for much water to collect in pores.) The separation between these two zones is called the groundwater table or *water table;* any hole with a depth below the water table will fill with water creeping in from the saturated zone, up to the level of the water table. In a manner of speaking, the water table is the surface of the underground water.

How far down is the water table? The water table tends to be deeper below hills than valleys, but to a large extent the depth depends on how much rainwater has recently infiltrated the surface. Warm, wet areas, such as parts of Florida in the United States, have a water table only about three feet (0.9 m) down. (If the water table reaches the surface, a spring appears.) But water tables in the desert are many times this far beneath the surface. Water is nearly always present if a driller is willing to dig far enough, but deep drilling is extremely expensive and pumping the water to the surface becomes an issue as well. In most climates, well diggers succeed when they find an area consisting of porous rock such as sandstone, which allows a lot of water to seep through and collect underneath.

Aquifers are vital to many communities, yet when people draw too much water too quickly from an underground source, it might not be replaced. As discussed in the sidebar on page 136, the rate that water is discharged or exits the aquifer must be balanced by the rate at which water recharges or flows into it. Otherwise, the well runs dry.

Groundwater is replenished by infiltration, but aquifer overuse due to increased demand causes water levels to fall dangerously low. All reservoirs—groundwater, lakes and rivers, glaciers and polar ice, and the oceans—have inflows and outflows that, when balanced, maintain a constant level. Imbalances arising from excessive withdrawal, seasonal variations, or climate change will exacerbate water scarcity and shortages. And although the planet's total amount of water is roughly constant—almost all water eventually returns to some reservoir or another—a significant portion may become polluted or end up in an inaccessible location, magnifying the cost and difficulty of quenching civilization's thirst.

FINDING AND USING FRESHWATER

Pollution has always been a serious threat to the freshwater supply of large cities. Waste disposal becomes a problem under crowded conditions, and sewage often finds its way into the city's streams, rivers, and groundwater. Bacterial contamination causes serious and deadly diseases such as cholera, which can spread rapidly. Civilizations as far back as ancient Rome have had to invest considerable time and expense into finding solutions to their water problems, which, in the case of the Romans, involved constructing a number of aqueducts to bring water into the city.

Many communities tackle the pollution problem by carefully routing their sanitation systems and by building sophisticated water purification and treatment systems. Water facilities disinfect and filter water that is piped into homes and businesses, removing most of the contaminants. But communities and villages in undeveloped areas of the world lack access to treated and purified water. A complete absence of water is not necessarily the problem in these communities—the absence of safe drinking water is what causes most of the trouble.

But sometimes rivers do run low and wells do go dry. These water shortages often occur during droughts, which constitute another major threat to water supplies. As described in the next section, prolonged dry spells are difficult to predict and can come without warning, catching officials by surprise and preventing the stockpiling of additional quantities of water. Droughts are common in arid regions such as deserts, but can strike anywhere. The following sidebar discusses the drought and its aftereffects that caused havoc in a large section of the United States in the 1930s.

Digging wells to reach new sources of water is often needed, but finding the best place to dig is not simple. One way to find out is to guess—well diggers have more success in rocks like sandstone, so this is a good place to start—and then drill a test hole. But guessing wrong too often is expensive.

The Dust Bowl of the 1930s

The United States fell on hard times in the 1930s after the stock market crashed and the Great Depression contracted the economy, leading to joblessness and poverty for millions. For the lower Midwest states, a second disaster struck at the same time—the dust bowl. A severe drought beginning in 1931 parched millions of acres of land in Texas, Oklahoma, Colorado, Kansas, and New Mexico. The surface turned to dust, and winds kicked up the dry soil into furious dust storms that blanketed much of the region in dark, gritty clouds. People who got caught outside in these storms had to grope their way home, even if they were close, since no light could penetrate the murk. Chickens came home for the night—in the middle of the day. On the PBS program *Surviving the Dust Bowl,* which aired in 1998, an eyewitness described the experience: "It kept getting darker and darker. And the old house is just a-vibrating like it was going to blow away. And I started trying to see my hand. And I kept bringing my hand up closer and closer and closer and closer. And I finally touched the end of my nose and I still couldn't see my hand. That's how black it was."

Bad farming practices contributed to the problem. In the years leading up to the 1930s, farmers had used the

Technology may soon be able to take some of the guesswork out of the process. Radar creates images of distant objects by bouncing electromagnetic radiation off them. Ground-penetrating radar (GPR) is a tool to image subsurface regions in a similar process, since properties of the soil, including its water content, affect the transmission of electromagnetic radiation. When a lot of water is present in the soil, the water tends to be more prominent than soil in determining the propagation of radiation. GPR has been used in oil exploration and land extensively with little thought of conservation. Repeated use of the same land for the same crop depleted the soil and led to serious erosion. Widespread plowing killed grass that trapped moisture and normally kept soil from blowing away. As a result, the drought parched the soil and the wind easily lifted it away, generating massive dust storms that people called Black Blizzard or Black Roller. The term *dust bowl* began to be used for the area after April 14, 1935, known as Black Sunday, which featured a large number of dust storms.

The consequences of the dust bowl have been long-lasting. People in the affected region suffered considerable health problems and economic hardships, and many people moved away. The migration out of the dust bowl states was the largest mass movement of people in U.S. history, with several million Okies loading up their trucks and heading for greener lands. (Although only about 20 percent of these migrants were from Oklahoma, the term *Okie* was generally applied.) Due in part to the depression, many migrants had trouble finding jobs, as dramatized in John Steinbeck's 1939 novel, *Grapes of Wrath*. One of the few beneficial aspects of the disaster, though, was the development and implementation of improved farming practices. A repeat of the dust bowl, even under severe drought conditions, is unlikely.

glacier measurements, and scientists such as soil surveyors in Fairfax County, Virginia, have used the technique to gauge soil layers and the location of water tables.

The difficulty with GPR is that it provides a clear image only under certain conditions and at shallow depths. Data from this technique is complicated and requires a great deal of analysis before the signals can be correctly interpreted. At best, the tool is a suggestive guide, and if the water table is deep, water-seekers can glean little useful information.

Animals like camels adapt to living in deserts and retain water in their digestive tracts. *(Sean Randall/Stockphoto)*

Until researchers concoct a superior method to detect subsurface water, people must rely on old-fashioned trial and error.

For people living near the coasts, finding water is as easy as going to the beach. The water is salty, though, and undrinkable. Yet it need not remain so. Enhancing the world's freshwater supply can be accomplished through *desalination*—the removal of salt from seawater.

Conversion from salt to freshwater is a natural process in the water cycle, as H_2O molecules evaporate from the ocean and later precipitate as freshwater. The conversion can be produced artificially by distillation or various other procedures. More than 100 countries worldwide use some amount of desalination, but in several countries of the Middle East, North Africa, and the Caribbean Islands, desalination is one of the most important sources of potable water. (More than half of the world's desalination facilities are located in the Middle East.) Less than 1 percent of freshwater in the United States comes from desalination processes, although in 2007 in Tampa Bay, Florida, a seawater distillation plant opened that can provide about 10 percent of the city's water supply when the plant is operating at full capacity.

The main disadvantage of desalination to produce freshwater is cost—the process is usually around four to five times more expensive than other water supplies because it uses a lot of energy. But for cities with little access to freshwater and abundant access to seawater, maintaining an adequate supply for its citizens is worth the extra cost.

Some countries could employ desalination to meet their needs but do not have the funds to construct high-capacity systems. However, researchers are working on developing low-cost alternatives. For example, Joachim Koschikowski at the Fraunhofer Institute for Solar Energy Systems in Freiburg, Germany, and his colleagues have recently built small desalination systems that use solar power. One of these systems can generate about 32 gallons (120 L) of freshwater per day at a cost of only a few dollars. Test generators in Jordan and at one of the Canary Islands in the Atlantic Ocean have been successful.

Another important factor is conservation. Since finding or generating freshwater is difficult and expensive and excessive pumping of groundwater leads to serious problems such as subsidence, consumers must make optimal use of what they have.

In addition to obvious strategies such as avoiding waste, scientific research and development will play an important role in optimizing water use. For example, greater efficiency in crop irrigation, which consumes a lot of water, would result in substantial savings. Biologists and soil scientists have discovered that maximal growth for crops depends on temperature and water, which varies from species to species. At the Agricultural Research Service, a research agency of the United States Department of Agriculture (USDA), scientists are working on water-saving feedback mechanisms, based on data obtained from plants, to control crop irrigation. The researchers Steven Evett, Susan O'Shaughnessy, and their colleagues at the Agricultural Research Service have recently attached sensors to crop plants; the sensors transmit information concerning plant temperature and health to the irrigation system. This information provides essential feedback to the irrigation system so that it can adjust the water supply as needed, depending on the crop status at any given moment. Such fine-tuning saves water because the system only delivers water when it is necessary.

Information fed back to a crop irrigation controller permits greater efficiency, but researchers would also like to be able to predict specific needs and supplies in advance. Shortages or excesses in all water use—industrial, agricultural, and personal—could be minimized if people knew about and could prepare for their future water situation. But to predict the future, hydrologists need a great deal of data and a little bit of luck.

HYDROLOGIC MODELING AND PREDICTION

Weather and climate are extremely important factors in a region's water supply. Rain fills lakes, swells rivers, and replenishes aquifers, while droughts diminish all of these water reservoirs. Hydrologists who develop models of the water cycle and the ebb and flow of supplies must take into account the weather. But predictions are difficult, and this is particularly true of weather.

The Massachusetts Institute of Technology (MIT) professor Edward Lorenz (1917–2008) discovered one of the reasons why weather prediction is so frustrating. In the 1960s, Lorenz was developing mathematical models to forecast weather systems. The model started with a set of initial conditions—temperature, wind speed, atmospheric pressure, and so forth—that Lorenz provided as an input. Then a set of equations computed the future course of weather starting from the initial conditions. But Lorenz noticed that the prediction was extremely sensitive to the exact values of the initial conditions. Even if the values changed only slightly, the model's behavior changed drastically, which meant that a small change in the values Lorenz inputted into the model resulted in a major change in the model's prediction.

Sensitivity to initial conditions is known as chaos. Chaos occurs in certain systems, such as weather systems, that are governed by complicated interactions. Predicting the future of these systems is exceptionally difficult because tiny changes in the present state of the system get magnified over time. The importance of small changes is sometimes called the butterfly effect, from the notion—perhaps only slightly exaggerated—that the flapping of a butterfly's wings in Texas can affect a hurricane in the Atlantic Ocean.

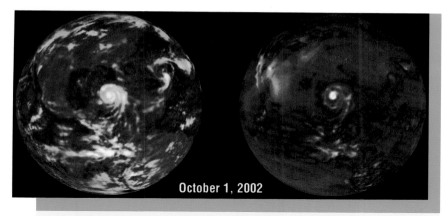

October 1, 2002

The image at left shows an *Aqua* satellite measurement of the amount of reflected light (due mostly to clouds) over the United States and the Gulf of Mexico. The image at right shows the amount of heat leaving the surface, with the red areas losing the most. Energy flow is critical to understanding the water cycle. Note the hurricane in the center of the images. *(CERES Science Team, NASA Langley Research Center)*

Values of the initial conditions are crucial, but there is always some uncertainty about these values because they come from measurements that can never be completely accurate. All measurements incur some amount of error, some of which is unpreventable—a yard or meter stick, for example, may provide accuracy only to the nearest eighth of an inch or to the nearest millimeter. As a consequence, weather models become wildly inaccurate after a few days.

To improve the accuracy of their models, hydrologists need to collect large amounts of accurate data. Because small changes at one location can influence distant events, the data needs to be global in scope. One of the best ways of gathering global data is from a position of great altitude. This requirement calls for a satellite.

The National Aeronautics and Space Administration (NASA), the U.S. government agency involved in space science and exploration, launches many satellites to study Earth from space. NASA's Earth Observing System consists of a series of satellites designed to keep an eye on the planet, and one of these satellites, *Aqua,* is specifically designed for hydrology. (The term *aqua* is Latin for "water.") *Aqua,* launched on

May 4, 2002, features six instruments to collect data on the water cycle, water vapor in the atmosphere, clouds, snow and ice, rainfall, soil moisture, and surface temperatures, among other variables. Data obtained from *Aqua* has helped researchers at the National Oceanic and Atmospheric Administration (NOAA) improve their forecasts, allowing better accuracy farther into the future.

Weather predictions of NOAA and other weather service organizations enable communities to plan ahead. Contingency plans for droughts are particularly important to preclude shortage crises. To meet this need, the U.S. government established the National Integrat-

National Integrated Drought Information System (NIDIS)

Widespread droughts in 2006 included many of the Great Plains states of the U.S. Midwest, and in 2007 droughts in the Great Plains as well as in the East cost more than $5 billion in agricultural and other losses. Severe water shortages are not rare—at any given time, nearly a third of the United States is affected. For example, in August 2007, nearly 40 percent of the country suffered from moderate to severe drought conditions (fortunately, the percentage dropped to 20 percent by May 2008). The West and Midwest—the old dust bowl region—experience droughts frequently, but other areas of the country are not immune.

Prompted by governors of western states and other officials, the United States established the National Integrated Drought Information System (NIDIS) in 2006. NIDIS is a collaboration of several government agencies, including the Department of Agriculture, Department of Commerce, Department of the Interior, NASA, and others, led by NOAA. The goal is to collect and integrate data from these agencies—satellite data, crop yields, water levels, and so forth—to assess and monitor drought conditions and develop increasingly accurate

ed Drought Information System (NIDIS). As described in the sidebar on page 146, NIDIS collects information from a spectrum of sources to monitor the current drought situation in the United States and to highlight any trends that may indicate what the future has in store.

Gathering the right data is essential, but hydrologists also need to understand how water moves and flows within its reservoirs. Vital to many locales is the flow of groundwater. How water moves through soil is critical to the maintenance of aquifers, but predicting this flow is difficult because soil is not generally the same throughout a given region—soil is usually heterogeneous, composed of a mixture of

models and forecasts. To spread their findings as widely as possible, NIDIS launched a Web site, www.drought.gov, on November 1, 2007. This Web site shows which parts of the country are presently affected by drought and which parts may be affected in the future.

Since weather plays a strong role in drought conditions and local weather can be influenced by remote events, drought forecasts are necessarily global in scope. For instance, an important factor in drought occurrences in the United States is El Niño, a periodic warming of the waters off the west coast of South America. (The name of this phenomenon derives from the Spanish term for "the little boy," a reference to Christ, as the phenomenon is often observed around Christmas.) El Niño and its corresponding atmospheric oscillation are associated with floods, storms, and droughts at a variety of locations around the world, which occur following particularly strong El Niño episodes. Researchers have found that El Niño is strongly linked to rainfall amounts along the coastal areas of the United States and weakly linked to the central areas of the country. La Niña—a cooling of the waters off the west coast of South America—seems much more important to the situation in the central portion of the United States. (La Niña is Spanish for "the little girl.") The nature of these links, and how they occur, are the subject of much ongoing research.

different types. Hans-Jörg Vogel of the Helmholtz Centre for Environmental Research—UFZ in Leipzig, Germany, and Olaf Ippisch at the University of Stuttgart in Germany recently refined models of these flows. Many of these models are based on a mathematical formula called Richards' equation, which is limited in how large an area it accurately models. At large scales—a large section of ground—the model must be broken up into discrete partitions of a certain size, otherwise it is inaccurate. Vogel and Ippisch found a way of estimating the size of these partitions so that the models would be correct. The researchers published their findings, "Estimation of a Critical Spatial Discretization Limit for Solving Richards' Equation at Large Scales," in a 2008 issue of *Vadose Zone Journal*.

Scientists are also monitoring aquifers to collect even more data. As crucial sources of water for many regions, aquifer depletion would have serious consequences. For example, the largest aquifer in North America, the Ogallala Aquifer, lies under parts of eight American states (Texas, New Mexico, Oklahoma, Colorado, Kansas, Nebraska, Wyoming, and South Dakota). A lot of farms and homes rely on this water. Ogallala's supply is dwindling, as estimated by the United States Geological Survey (USGS), and although it is continually recharged, replenishment happens slowly and over a limited area. Dennis Gitz of the Agricultural Research Service and his colleagues at Texas Tech University are monitoring the flow of water through the soil around the aquifer with soil thermometers (the presence of water alters the soil's temperature). The researchers are focusing on playa lakes—temporary lakes formed when rainwater collects in a cavity—to see if water filtering through the soil at these points is contributing much clean water to the aquifer. If so, then the playa lake region must be maintained and protected. Gitz and his colleagues have begun the study by installing sensors at 14 playa lakes and are preparing to complete 16 others.

As the quality of data improves, so will hydrologic models and predictions. Yet researchers may find themselves trying to hit a moving target—any modification in the climate affects the water situation, and the world's climate seems to be in the midst of substantial changes.

CLIMATE CHANGE AND WATER

Global warming has not been uniform. Some regions, such as the southeastern United States, have cooled slightly during this time, and some regions, such as parts of Canada and northern Europe, have warmed at twice the average rate.

Scientists—as well as everybody else—would very much like to know what is causing global warming. An important contributor is emissions from factories, automobiles, and other human activities that have increased the amount of greenhouse gases such as carbon dioxide in Earth's atmosphere. These gases tend to raise temperatures by absorbing infrared radiation, thereby trapping heat. Attributing most of the recent warming trend to greenhouse gas emissions is a reasonable hypothesis, and many scientists accept it, although it is difficult to prove. Previous warming trends in Earth's history, such as the one that ended the last of the ice ages about 12,000 years ago, have occurred well before human industry arose. No one is certain what the future climate will be like—Lorenz showed how predictions of complex phenomena such as weather and climate are usually erroneous.

How will global climate change affect the planet's hydrology? Global averages of precipitation have not changed much over the last century, although there has been variability—some tropical and equatorial regions have experienced less rainfall than usual and other latitudes have had more. But the warming trend has begun to melt a significant amount of ice on and around the polar regions. NASA studies indicate that the Arctic ice thickness has diminished about 40 percent in the last few decades, and glaciers in Greenland and Antarctica are retreating. Losses of sea ice—a thin layer of ice over water—have been severe, with an area of sea ice the size of Norway, Denmark, and Sweden combined having vanished from the Arctic region.

The consequences of melting glaciers will be rising sea levels. As water shifts out of the ice reservoir, much of it will end up in the oceans. The additional water will creep up the shores of continents and islands, flooding low-lying areas.

Other impacts of global climate change on the water cycle are less certain. Periodic changes in the properties of oceans, such as the warming of El Niño and the cooling of La Niña in the central Pacific Ocean, correlate with droughts or storms in other parts of the world, even in distant regions such as the United States. Due to the butterfly effect, nearly any change anywhere in the globe can exert some degree of influence on any other region.

In order to gather clues on what to expect in the future, some scientists are studying the past. For example, searching for the cause of episodes of extreme weather that have occurred in the past may give some indication of the future course of events. One of the most disruptive episodes in terms of water is a drought. Perhaps the best-known drought in the United States and the one with the greatest impact on American history was the long-lasting drought associated with the dust bowl.

Most scientists use models to study phenomena that take place on a global scale. Models, such as those describing Earth's interior, as discussed in chapter 1, or Earth's magnetic field, as discussed in chapter 2, distill what researchers believe is the essence—the critical features—of the problem into a simplified set of equations or structures. If the researchers have correctly identified the essential features, the model reflects the behavior and properties of the phenomenon. If not, the model is misleading.

Siegfried D. Schubert, a researcher at NASA's Goddard Space Flight Center in Greenbelt, Maryland, and his colleagues constructed a climate model based on historical data of sea surface temperatures in the 20th century. The researchers also used another model, developed at NASA, involving the atmosphere and its general circulation, the features of which came from observations obtained with satellites such as *Aqua* of clouds and precipitation patterns. A powerful computer simulated the behavior of the models and the time course of the weather patterns and temperatures by solving the various equations and crunching the data. With these tools, Schubert and his colleagues focused on the relation between sea surface temperatures and rainfall in the Great Plains states in the 1930s.

El Niño could have played a role in the 1930s drought, and fluctuations in the sea surface temperature in the Pacific Ocean did occur during the 1930s. But these fluctuations were mild and seem insufficient to account for the prolonged drought conditions during the dust bowl. What Schubert and his colleagues discovered was that a slight cooling of tropical Pacific Ocean temperatures coincided with unusually warm tropical Atlantic Ocean temperatures, and this altered the positions of high-velocity winds in the atmosphere. These winds have a significant affect on temperatures, as they guide or block the movement of air masses.

Atmospheric winds can also play a strong role in precipitation. Schubert and his colleagues found that shifts in ocean temperature during the 1930s altered the flow of a wind system that normally picks up moisture from the Gulf of Mexico. Under typical conditions, this moist air travels over the United States, particularly the Great Plains states, where it cools and falls as rain. Without this moisture, the Great Plains dried up, and the conditions affecting the wind lasted for an extended period of time, resulting in the devastating length of the 1930s drought. Schubert and his colleagues published their findings, "On the Cause of the 1930s Dust Bowl," in a 2004 issue of *Science.*

What will global warming, the loss of polar ice, rising sea levels, and other climate changes have on the water cycle and water supplies? Some models suggest a plausible scenario in which the warming trend will result in increased evaporation, which in turn will lead to more precipitation. This would be good news, at least for the reduction in the number and severity of droughts. But a NASA study suggests that the outlook is not necessarily good in terms of precipitation. Michael G. Bosilovich, Schubert, and Gregory K. Walker of the Goddard Space Flight Center used the atmospheric model mentioned above to examine what may happen to the water cycle. Their model also suggests higher precipitation levels, but the increase is over water, not land.

While higher temperatures increase evaporation, the warmer air can also hold more water vapor. The model of Bosilovich and his colleagues predicted higher cycling rates over water than land in general. In other words, the greater evaporation from the seas also fell on the seas in a rapid water cycle, while on land the opposite was true. The researchers published their report, "Global Changes of the Water Cycle Intensity," in a 2005 issue of the *Journal of Climate.*

No one can be sure at this point what the future will hold, but researchers need to continue to improve their models. As the University of Tokyo researchers Taikan Oki and Shinjiro Kanae wrote in *Science* in August 25, 2006, "Any change in the hydrological cycle will demand changes in water resource management, whether the change is caused by global warming or cooling, or by anthropogenic or natural factors. If society is not well prepared for such changes and fails to monitor variations in the hydrological cycle, large numbers of people run the risk of living under water stress or seeing their livelihoods devastated by water-related hazards such as floods."

CONCLUSION

Uncertainties in the future of Earth's water supplies are mirrored in the uncertainties and gaps in the scientific understanding of the water cycle. Most of the world's water is salty and undrinkable without desalination, which is an expensive procedure. Burgeoning populations, along with a rise in pollution, may result in unsustainable demands on freshwater sources such as rivers and aquifers. Innovations to increase water efficiency help ease the burden, but conservation and management of water sources are imperative.

The extent to which conservation and management must go to protect these water resources depends on the effects climate change may exert. If disruptions in weather patterns cause an increase in the number of areas experiencing prolonged drought or storms and flooding, strict measures may have to be taken. These measures may include restrictions on supplies and use, which in certain parts of the world must already be instituted from time to time. For example, during water shortages experienced in 2008, residents of Cyprus—an island nation located in the eastern Mediterranean Sea that has been averaging only 18.4 inches (46 cm) of rain a year for the last three decades—had their water cut off on certain days in order to ration the meager supply. Using a water hose for washing patios or cars was prohibited.

A better understanding of large-scale phenomena such as the world's water cycle requires extensive observations. A model running on a computer can simulate global weather patterns and predict what the future may entail, but the predictions will invariably be wrong unless the data and conditions used in the simulation are highly accurate.

To make observations on a worldwide scale, the best tool is a satellite. Orbiting high above the planet, sensitive instruments on board the satellite can watch over vast swaths of land, water, and atmosphere. *Aqua* and similar satellites have been useful, but more satellites are needed. NASA announced in the spring of 2008 that it plans to launch a satellite in December 2012 to map soil moisture. Scientists presently do not have any means of monitoring soil moisture globally, so they have to rely on samples taken at a few scattered points. Soil moisture has strong effects on evaporation and the water cycle and is a key feature in the cycling of carbon (organic material) and stored energy. An 19.7-foot (6-m) antenna will survey areas 620 miles (1,000 km) wide at a time and examine the entire globe every few days. Worldwide measurements of soil moisture will greatly aid climate and hydrologic models.

This data, along with fast computers and the skill and knowledge of researchers, will improve the accuracy of weather and water cycle models. Although the butterfly effect remains a serious impediment, the advances in modeling will reduce uncertainty and narrow the range of possible outcomes predicted by the models. This research is much needed. A University of Illinois researcher Mark A. Shannon and his colleagues issued a warning in *Nature* in March 20, 2008: "In the coming decades, water scarcity may be a watchword that prompts action ranging from wholesale population migration to war, unless new ways to supply clean water are found." With new satellite data and improved prediction techniques, scientists and government officials may be able to make well-informed decisions to manage, conserve, and replenish existing water supplies with minimal disruption to society.

CHRONOLOGY

312 B.C.E.	Romans begin building aqueducts to carry freshwater into the city.
1911 C.E.	Americans begin tapping the Ogallala Aquifer, the largest aquifer in North America.
1928	Curaçao, an island in the Caribbean Sea, constructs a desalination facility, one of the first major investments in desalination technology.
1930s	The worst drought to strike the United States affects much of the nation, but particularly an area in the Great Plains states of Texas, Oklahoma, Colorado, Kansas, and New Mexico. Drying of the soil, coupled with poor land management, results in severe dust storms that blanket the dust bowl region.
1950s	After strong episodes of El Niño, researchers begin to link this phenomenon with storms and droughts in the United States and elsewhere.
1960s	The MIT professor Edward Lorenz (1917–2008) discovers the butterfly effect—small changes in weather systems can have enormous consequences.

1974	The U.S. government passes the Safe Drinking Water Act, which regulates water treatment and sets appropriate standards.
2002	NASA launches the *Aqua* satellite. The collected data improves weather forecasts and hydrologic modeling and prediction.
2003	The UN issues its first World Water Development Report, warning of impending shortages, and designates the years 2005–2015 as the Water for Life Decade, urging conservation and careful management of water resources.
2006	In response to serious water shortages, especially in the western states, the United States establishes NIDIS to coordinate water monitoring and research efforts across the country.
2007	Tampa Bay desalination plant begins operations. When operating at full capacity, the plant can supply about 10 percent of the city's freshwater needs.
2008	NASA announces a tentative launch date of 2012 for a satellite designed to measure soil moisture.

FURTHER RESOURCES
Print and Internet

Bosilovich, Michael G., Siegfried D. Schubert, and Gregory K. Walker. "Global Changes of the Water Cycle Intensity." *Journal of Climate* 18 (2005): 1,591–1,608. The researchers' model predicts that global warming will lead to higher rainfall, but not on land.

Egan, Timothy. *The Worst Hard Time: The Untold Story of Those Who Survived the Great American Dust Bowl.* New York: Mariner Books, 2006. This history of the 1930s dust bowl describes the economic, ecological, and human catastrophe in vivid detail.

Environmental Protection Agency. "Water." Available online. URL: http://www.epa.gov/ebtpages/water.html. Accessed May 4, 2009. The EPA's mission is to monitor and protect the environment of the United States and the health of its citizens. The safety of drinking water is extremely important, and this Web resource discusses the problems posed by various sources of pollution.

National Aeronautics and Space Administration. "Aqua." Available online. URL: http://aqua.nasa.gov/. Accessed May 4, 2009. This Web resource describes the *Aqua* satellite, its instruments, the mission, and some of the results and images from the craft.

Oki, Taikan, and Shinjiro Kanae. "Global Hydrological Cycles and World Water Resources." *Science* 313 (August 25, 2006): 1,068–1,072. Oki and Kanae discuss freshwater resources and how water cycles affect their quantity and availability.

Outwater, Alice. *Water: A Natural History.* New York: Basic Books, 1996. Water is constantly on the go. This book eloquently describes the journey, from lake to house drain and back again, as water travels through complex ecological systems.

Pearce, Fred. *When the Rivers Run Dry: Water—The Defining Crisis of the Twenty-First Century.* Boston: Beacon Press, 2006. There is always a temptation to sensationalize any of the world's problems into a crisis for the sake of expanded news coverage, book sales, and so forth. But water is vital to life, and freshwater resources are becoming increasingly scarce, as the author cogently discusses in this book.

Postel, Sandra. *Pillar of Sand: Can the Irrigation Miracle Last?* New York: W. W. Norton & Company, 1999. Crop irrigation requires a significant portion of today's freshwater resources, and for thousands of years irrigation has played a critical role in boosting agriculture and meeting civilization's growing food demands. Water shortages imperil this process, but innovations and greater efficiencies offer hope for continued success.

Public Broadcasting Service. "Surviving the Dust Bowl." Available online. URL: http://www.pbs.org/wgbh/amex/dustbowl/. Accessed May 4, 2009. The Internet companion to an episode of *American Experience,* these pages include a time line of the events and interviews with eyewitnesses.

Schubert, Siegfried D., Max J. Suarez, et al. "On the Cause of the 1930s Dust Bowl." *Science* 303 (March 19, 2004): 1,855–1,859. The researchers develop a climate model that may explain the cause of the 1930s dust bowl.

ScienceDaily. "How Will North America's Largest Aquifer, the Ogallala Aquifer, Fare?" Available online. URL: http://www.sciencedaily.com/releases/2008/04/080405094350.htm. Accessed May 4, 2009. Dennis Gitz of the Agricultural Research Service and his colleagues at Texas Tech University are monitoring the flow of water through the soil around the Ogallala Aquifer with soil thermometers.

———. "Precision Irrigation Built into Sprinkler Booms Controls Water Usage, Optimizes Crop Growth." Available online. URL: http://www.sciencedaily.com/releases/2008/04/080420111817.htm. Accessed May 4, 2009. Steven Evett, Susan O'Shaughnessy, and their colleagues at the Agricultural Research Service are using sensors attached to crop plants to transmit information concerning plant temperature and health to the irrigation system.

Shannon, Mark A., Paul W. Bohn, et al. "Science and Technology for Water Purification in the Coming Decades." *Nature* 452 (March 20, 2008): 301–310. The researchers review the progress and future problems of water purification technology.

Tampa Bay Water. "Desalination Plant Fully Operational." Available online. URL: http://www.tampabaywater.org/watersupply/tbdesal.aspx. Accessed May 4, 2009. Tampa Bay Water describes their desalination plant, which began operating in December 2007.

Texas Council for the Humanities Resource Center. "The Dust Bowl." Available online. URL: http://www.humanities-interactive.org/texas/dustbowl/. Accessed May 4, 2009. The hardships of life in the dust bowl are highlighted, including many photographs and an essay.

United States Geological Survey. "The Water Cycle." Available online. URL: http://ga.water.usgs.gov/edu/watercycle.html. Accessed May 4, 2009. With many diagrams and photographs, this Web resource explains how the water cycle works. Topics include groundwater discharge and storage, runoff, infiltration, precipitation, springs, water vapor in the atmosphere, evaporation, and many others.

————. "Water Resources of the United States." Available online. URL: http://water.usgs.gov/. Accessed May 4, 2009. Maps, annual water reports, regional studies, and monitoring data are included in these extremely informative pages.

de Villiers, Marq. *Water: The Fate of Our Most Precious Resource.* New York: Mariner Books, 2001. Earth's rising population puts added demands on water resources, and people have not always managed these resources wisely. This book discusses water use from a historical, ecological, cultural, and political perspective. Topics include the distribution of water, climates, dams, aquifers, and irrigation.

Vogel, Hans-Jörg, and Olaf Ippisch. "Estimation of a Critical Spatial Discretization Limit for Solving Richards' Equation at Large Scales." *Vadose Zone Journal* 7 (2008): 112–114. The researchers present a refined model of groundwater flow.

Web Sites

National Integrated Drought Information System. Available online. URL: http://www.drought.gov. Accessed May 4, 2009. The NIDIS Web site offers maps and information showing which parts of the United States are currently experiencing a drought and how long it might last.

National Oceanic and Atmospheric Administration. Available online. URL: http://www.noaa.gov/. Accessed May 4, 2009. A wealth of information is available at NOAA's home page, including weather forecasts and climate research.

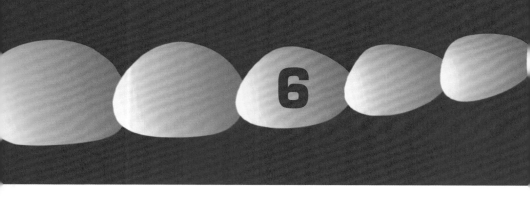

PREDICTING
EARTHQUAKES

At 2:28 in the afternoon of Monday, May 12, 2008, millions of Chinese in Sichuan Province and the surrounding area felt the ground start to shake. Sichuan Province is a populous region of China and home to many farmers as well as businesses. Many of the homes, factories, schools, bridges, and roads could not withstand the violent shaking. In the devastation that followed, more than 69,000 people lost their lives, several hundred thousand suffered injuries, and 5 million were left without homes—all in the space of a few minutes. This terrifying episode was an earthquake ("quake" derives from the Old English word *cwacian,* meaning "to shake or tremble"). *The Guardian* (Manchester) reported on May 13 that the earthquake and its aftereffects "caused panic and mass evacuations in cities across the country, including Beijing, 930 miles away, Shanghai and Wuhan. They were felt as far away as Vietnam and Thailand, 1,300 miles to the south. In Shanghai, China's financial centre, skyscrapers swayed as the tremor hit, sending office workers rushing into the streets."

The tragedy was not a novel one for the Chinese. A large number of earthquakes have struck China in the past, including one in the city of Tangshan on July 28, 1976, which killed about 250,000 people, making this event one of the deadliest disasters of the 20th century. Earthquakes also cluster in other regions of Asia; on December 26, 2004, an undersea earthquake in the Indian Ocean generated a huge wave known as a tsunami—a Japanese term for harbor wave—that swept over low-lying areas in Indonesia and neighboring regions, killing more than 250,000 people. California has also experi-

158

enced many earthquakes, including a San Francisco earthquake on April 18, 1906, that destroyed the city and claimed about 3,000 lives.

Most of the damage and casualties from earthquakes are due to collapsing structures or scattered debris. The ground shakes or oscillates because of earthquake waves, or seismic waves, which spread out from the earthquake's origin—the focus (also known as the hypocenter)— and travel in all directions. Many communities that have experienced numerous earthquakes require builders to follow strict codes. Buildings and bridges can be designed to resist at least a moderate amount of

This California highway overpass collapsed during a 1971 earthquake. (R. Kachadoorian/USGS)

shaking, though there is still danger from flying objects and buckling floors, and a powerful earthquake can level almost any structure. The safest strategy would be to evacuate cities and vulnerable buildings before the earthquake starts—if advance warning could be provided.

Because the properties of Earth's interior affect the properties of seismic waves, geologists have been studying these waves since the 19th century to learn something about the planet's inner structure, as discussed in chapter 1. Seismic waves are also crucial pieces of information concerning their source—earthquakes. Yet after more than a century of study, scientists have been unable to predict when and where an earthquake will occur. The United States Geological Survey (USGS), California Geological Survey, and Southern California Earthquake Center released a report in 2008 saying that California has a greater than 99 percent chance of suffering a major earthquake within 30 years. According to the report, a strong earthquake "is virtually assured in California during the next 30 years."

But the report does not specify exactly when the earthquake will occur or what part of California will be hit. This ambiguity limits the report's usefulness. Researchers at the frontier of Earth science would like to do better. Earthquake forecasts often rely on historical records and the tendency of earthquakes to recur in certain areas. This chapter discusses ambitious research projects that aim to use techniques such as animal behavior, tremors, and fault monitoring to improve earthquake forecasts.

INTRODUCTION

Geologists gained an important clue about the cause of volcanic activity when they realized that the vast majority of volcanoes are clustered around the boundaries of tectonic plates. The same is true for earthquakes. An additional link is that most of the major earthquakes occur around the Ring of Fire, the narrow ribbon of volcanic and seismic activity that encircles the Pacific Ocean (see the figure on page 74). The movement of these massive plates is clearly associated with both earthquakes and volcanoes.

In the 19th century, about 100 years before scientists became aware of tectonic plates, geologists began fashioning instruments called seismometers to record seismic waves. A seismometer is a device that measures the shaking of the ground as a seismic wave passes by. In the simplest case, the instrument consists of a freely moving

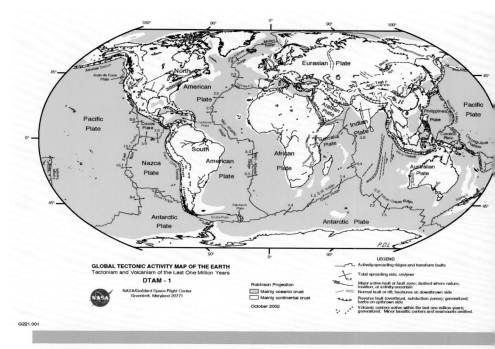

GLOBAL TECTONIC ACTIVITY MAP OF THE EARTH
Tectonism and Volcanism of the Last One Million Years
DTAM - 1

NASA/Goddard Space Flight Center
Greenbelt, Maryland 20771

Robinson Projection

October 2002

Mainly oceanic crust
Mainly continental crust

LEGEND

Actively-spreading ridges and transform faults

Total spreading rate, cm/year

Major active fault or fault zone; dashed where nature, location, or activity uncertain

Normal fault or rift; hachures on downthrown side

Reverse fault (overthrust, subduction zones); generalized; barbs on upthrown side

Volcanic centers active within the last one million years; generalized. Minor basaltic centers and seamounts omitted.

G221.001

This map depicts plate tectonic activity over the past 1 million years. *(NASA)*

weight, such as a small block, attached to a spring or a hinge. As the wave passes, the block swings back and forth. A pen or writing instrument attached to the block makes a mark on a roll of paper as it glides past—a process called seismography—and records the motion over the course of time on a seismogram. The seismograms of these pioneering researchers enabled them to discover the structure of Earth's interior, including the inner core, outer core, and mantle. Seismologists of today use electrical devices to magnify the instruments' motion, enhancing the sensitivity of seismometers so that they can detect displacements as small as the diameter of a molecule!

Seismic waves emanate from the earthquake's focus, as shown in the following figure, eventually reaching distant seismograph stations. These waves vary in frequency from about 0.1 to 30 hertz (cycles per second). A lot of the energy of seismic waves dissipates as it travels through rocks and soil, which decrease the magnitude of the waves. The area directly above the focus is known as the *epicenter*. Most earthquakes originate less than about 50 miles (80 km) below the surface, so the epi-

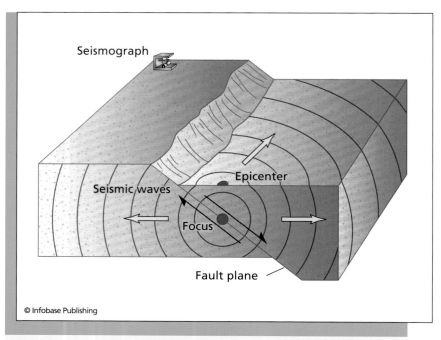

Seismic waves propagate in all directions from the earthquake's focus.

center is quite near the focus and is usually the hardest hit region in an earthquake. Seismologists can calculate the location of an earthquake's focus by studying the time of arrival of the seismic waves of various types (see chapter 1).

Some earthquakes shake the ground a lot, but other earthquakes create only a small disturbance. The duration of an earthquake can be a few seconds to a few minutes, but people feel the ground shaking for about 30 seconds in an average earthquake. In 1902 the Italian researcher Giuseppe Mercalli (1850–1914) proposed a scale to measure earthquake intensity based on observational evidence. An intensity of 1 on this scale was detectable only with seismometers, while 12 equated total destruction. Slightly modified in 1931, this scale is still used occasionally to describe the severity of an earthquake's effect. The modified Mercalli scale uses Roman numbers from I to XII, with I being least and XII maximal. A moderate earthquake is a IV—cars can be seen rocking and dishes rattle—and a very strong one is a VII, resulting in slight to moderate damage to a typical building.

However, since observations are not always reliable and damage depends on the engineering properties of the affected structures, a more precise scale is needed. (Bigger earthquakes usually cause much more damage, although the amount of damage also depends on the fragility of the buildings and structures affected by the motion.) To compare the magnitudes of different earthquakes, scientists used to rely solely on seismic wave recordings. In 1935 Charles F. Richter (1900–85), a geologist at the California Institute of Technology, and his colleague Beno Gutenberg (1889–1960) developed a scale to measure the size of an earthquake. As described in the following sidebar, the Richter scale (sometimes called the Richter-Gutenberg scale) assigns a number based on the logarithm of the amplitude of an earthquake's seismic waves. The amplitude of a wave is the size of its peak or maximum deviation from zero or flat line.

Because the Richter scale uses base-10 logarithms, the vibrations of, say, a 5.0 earthquake, are 10 times greater than those of a 4.0 earthquake. The seismic waves of a 2.0 earthquake on the Richter scale carry roughly the equivalent amount of energy as a ton of TNT. The seismic waves of a 5.0 carry about the same energy as the nuclear weapon detonated over Nagasaki, Japan, on August 9, 1945, and the seismic waves of a 7.0 have as much energy as that released by the largest nuclear bomb ever tested—Russia's Tsar Bomba—equivalent to 50,000,000 tons of TNT.

Until recently, everyone used the Richter scale. But now seismologists employ a more direct measurement of earthquake intensity called *moment magnitude,* which is based on the actual movement that causes the earthquake, as described below. This scale generates values similar to the Richter scale, but a moment magnitude is usually more representative of the earthquake's energy. According to the USGS Earthquake Magnitude Policy, "Moment magnitude is the preferred magnitude for all earthquakes listed in USGS catalogs."

The largest recorded earthquake occurred in Chile in 1960 and registered 9.0 on the Richter scale and 9.5 on the moment magnitude scale, which is abbreviated M_w. Other sizable events include an Alaska earthquake in 1964 that registered 8.4 on the Richter scale and 9.2 M_w (the largest ever recorded in the United States). Geologists estimate that the 1906 San Francisco earthquake was approximately 8.0 on the Richter scale. Most people now report the moment magnitude as the earthquake's magnitude without mentioning they are not using the Richter scale, which confuses readers. Seismologists reported that the Sichuan earthquake of

Richter Scale—an Early Method of Quantifying Earthquake Intensity

The California Institute of Technology researchers Charles Richter and Beno Gutenberg developed the scale in accordance with the amplitude of the vibrations as measured by their particular seismometer. By comparing instrument readings rather than subjective observations, the researchers could judge the size of any earthquake. This method permitted them to distinguish between the numerous smaller earthquakes and the rare but important major ones without having to rely on eyewitnesses.

The range of seismic amplitudes is large—the amplitudes of some seismic waves are huge compared to others. A scale with such a wide range is unwieldy because it must include enormous numbers as well as tiny ones. To make the numbers more manageable, Richter assigned earthquake magnitudes based on the logarithm of the amplitude. Logarithms compress the range; for example, the logarithm (base 10) of 10 is 1, and

May 12, 2008, measured 7.9 M_w (not Richter!). The magnitude of the undersea earthquake that generated the December 26, 2004, Indian Ocean tsunami was 9.3 M_w.

A large number of earthquakes occur every year, though most are fortunately so small that only sensitive instruments can detect them. About 1,000,000 earthquakes above magnitude 2.0 M_w occur a year. (If one considers any movement, no matter how small, to be an earthquake, then earthquakes are continuous—the ground is in motion at some point on Earth at almost any given time.) Only about a tenth of these earthquakes exceed 3.0 M_w, which is the smallest a person can usually feel. Ten earthquakes on average will exceed 7.0 M_w. The state with the most earthquakes in the United States is Alaska (California is second).

Even before scientists knew about tectonic plates, they guessed the basic mechanism of earthquakes. In 1760, British scientist John Michell the logarithm of 100 is 2. (The base-10 logarithm y of a number x is given by the formula $10^y = x$.) Richter chose the scale's 0 value to be a certain extremely small amplitude as recorded by his instrument when located 62 miles (100 km) from the epicenter. An amplitude 10 times greater than this value would register 1 on the Richter scale, 100 times greater would register 2, 1,000 times greater would register 3, and so on.

A seismic wave's amplitude depends on the distance from the focus as well as the sensitivity of the recording instrument, but a mathematical scale was so useful that scientists adapted the Richter scale for a variety of instruments and distances. In each case, seismologists calibrate the output of their instrument to achieve consistent readings of the earthquake's intensity that would be observed 62 miles (100 km) from the epicenter. There is no minimum or maximum on the scale. Although 0 is an extremely small value on the Richter scale—it was about the least that the old instruments could measure—newer instruments are sensitive enough to detect smaller amplitudes, which measure negative values on the Richter scale.

(1724–93) proposed that rock movements deep below the surface cause earthquakes to happen. But it took a while before scientists realized how and why these rocks are moving.

FAULT ZONES

Alfred Wegener (1880–1930) proposed a theory of continental drift in 1912, and although his ideas were not entirely correct, geologists in the 1960s realized Earth's crust is composed of about 12 rigid plates and a few dozen smaller ones, all moving and jostling each other. These tectonic plates ride on a partially molten layer called the asthenosphere. The motion is slow, about 1–6 inches (2.5–15 cm) per year on average, but has dramatic effects. Plate boundaries create space where magma oozes to the surface, creating the majority of Earth's volcanoes. And as plates bump, slip underneath, or grind past one another, tremendous forces are un-leashed. These forces are responsible for most of the planet's earthquakes.

A fault is a crack or fissure in which one side or wall moves rela-tive to the other. Some faults are short, but others extend for 100 miles (160 km) or more. Faults often occur around plate boundaries, and the

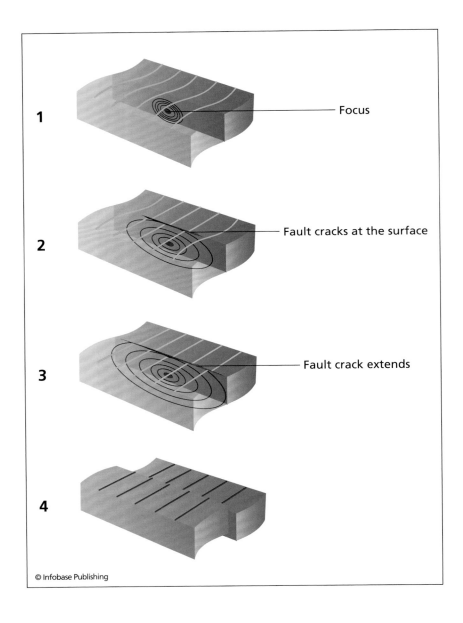

1 — Focus

2 — Fault cracks at the surface

3 — Fault crack extends

4

© Infobase Publishing

motion of the plates create the forces that move the rocks. As the rocks on one side try to slide past the rocks on the other, they may get stuck, arresting the motion temporarily. Stress builds in the fault, as the forces continue to push or pull against the rocks. The rocks exhibit strain, deforming or bending, due to the tremendous forces, and finally they break. A break may occur along an existing fault or it may open up a new crack, but, either way, the result is a sudden movement and the release of a huge quantity of energy. These events cause earthquakes. The figure at left illustrates the process.

Most plate boundaries have complicated geometries and do not consist of a single fault running the length of the boundary. Instead, a number of different faults exist, creating a fault zone or system. California contains a major fault system that has been responsible for numerous earthquakes in the region. One of the most destructive faults in this system is the San Andreas Fault, described in the following sidebar.

Following the 1906 San Francisco earthquake, geologists studied the San Andreas Fault extensively. They discovered that the giant earthquake had been caused by a sudden, massive shift along the fault, as evidenced by offsets in roads and other structures that crossed the fault. Although the theory of plate tectonics was still in the future, in 1910 Henry F. Reid (1859–1944), a geologist at Johns Hopkins University in Baltimore, Maryland, realized how earthquakes occur. After studying the San Andreas Fault and the 1906 San Francisco earthquake, Reid proposed a theory of earthquakes called elastic rebound theory. An elastic material, if deformed by some force, will snap back into place after the force disappears; for example, a finger pressed into an elastic rubber ball will deform the ball's shape, but after the finger lifts, the ball will regain its spherical shape. Reid believed that stress builds up over time along a fault, deforming the rocks, until something finally gives and the rocks snap back, or rebound, causing a slip along the fault. (See the figure on page 166.) This sudden movement by a massive amount of rock sends out a huge amount of shock waves—an earthquake.

(opposite page) Strain develops along a fault, eventually producing a rupture (1), which quickly reaches the surface (2) and spreads (3), extending throughout the fault (4).

San Andreas Fault

Two large tectonic plates, the Pacific plate and the North American plate, meet in California. Part of the boundary includes the San Andreas Fault, as shown in the following figure. The San Andreas Fault takes its name from San Andreas Lake, which lies a little south of San Francisco in a valley created by the fault. Andrew Lawson (1861–1952), a professor at the University of California, Berkeley, identified the northern stretch of the fault in 1895 and later discovered it extended far to the south. San Andreas is the backbone or master fault in the system, running about 800 miles (1,280 km) from northern California to San Bernardino in the south. The fissure extends to a depth of at least 10 miles (16 km).

Rocks on opposite sides of the San Andreas Fault move past one another horizontally. This motion is due to the plate movement—the Pacific plate moves northwestward with respect to the North American plate at a rate of about 2 inches (5 cm) per year, as measured in the San Francisco area. Since Los Angeles is on the Pacific plate and San Francisco is on the North American plate, the two cities will slide past each other in a few million years if the plates continue their present motion!

In 1906, only 11 years after Lawson's discovery, this fault became the center of much attention from geologists—it was the origin of the tragic San Francisco earthquake. Geologists who flocked to the site found that fences, streams, and roads that stretched across the fault were no longer lined up, for one side had suddenly shifted. Instead

(continues)

Fault slips are the basis for the moment magnitude measurements mentioned above. The product of the area of a fault's surface and the average distance it moved during a slip is called the moment

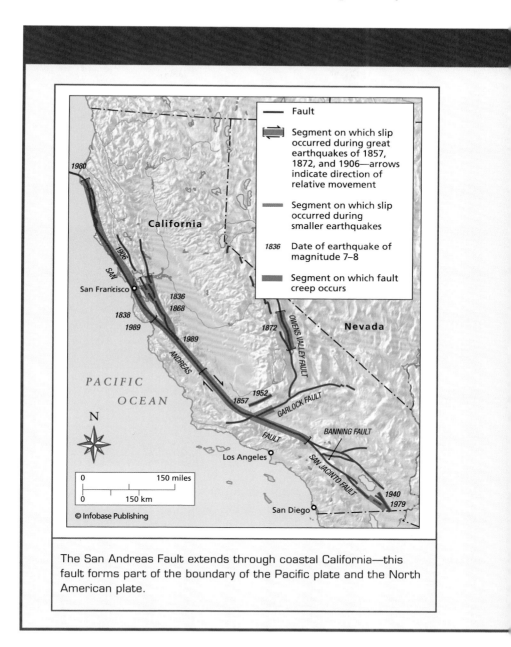

—	Fault
⬛➡	Segment on which slip occurred during great earthquakes of 1857, 1872, and 1906—arrows indicate direction of relative movement
▬	Segment on which slip occurred during smaller earthquakes
1836	Date of earthquake of magnitude 7–8
▬	Segment on which fault creep occurs

The San Andreas Fault extends through coastal California—this fault forms part of the boundary of the Pacific plate and the North American plate.

of an earthquake. Scientists can estimate the moment from seismograms, but they can also determine the moment by studying the fault itself.

(continued)

This view shows a portion of the San Andreas Fault at the Carrizo Plain. The fault runs horizontally across the middle of the photograph. Note that the stream channel running vertically is out of alignment due to movement along the fault. *(R. E. Wallace/USGS)*

of a continuous road or fence, one side was offset from the other side, across the fault. In some cases, such as the road at Tomales Bay, the offset was nearly 21 feet (6.4 m)—the center of the road at one side of the fault was a horizontal distance of 21 feet (6.4 m) from the road at the other side of the fault!

A few earthquakes occur far from any plate boundary, similar to the phenomenon of volcano hot spots. The origin of these earthquakes is not generally well understood. But cracks or faults within plates would explain why these events occur, and in some cases evidence for these faults has been discovered.

Thanks to the work of Lawson, Reid, and numerous other researchers, geologists now have a good idea how and why earthquakes occur. But earthquakes are complex. Geologists can easily identify faults and specify which regions are likely to experience major earthquakes in the future, but precise predictions have proven difficult. The large size of the plates and the complicated nature and geometry of their interactions have thus far defied specific predictions of future events. Determining if an earthquake will happen at a given place tomorrow or next week is not yet possible, since the uncertainty is given in decades or even centuries.

But Earth scientists are continuing to work on earthquake prediction, and their motivation is not just scientific curiosity. The May 12, 2008, earthquake in Sichuan Province, China, killed tens of thousands of people and is not a rare event. Major earthquakes occur every few years, causing hundreds or thousands of deaths and billions of dollars in damage. A number of populous cities or regions are threatened, including San Francisco, Los Angeles, Tokyo, Tehran, Istanbul, Mexico City, and many others. At the very least, researchers want to develop warning systems that would give people a chance to seek safety before the seismic waves arrive and the buildings start to crumble.

WARNING SYSTEMS

A warning system is not the same as the ability to make a prediction. Suppose that 36 miles (58 km) away from city C, a fault slips, creating dangerous seismic waves. As chapter 1 described, the fastest seismic waves, the primary or P waves, travel through rock at about 13,000 miles per hour (20,800 km/hr), or 3.6 miles (5.8 km) a second. The waves take about 10 seconds to reach C. If the city had a sensor to detect this event and send a message that traveled much faster than the seismic waves—for instance, by radio, which is 50,000 times faster—then the citizens would be alerted a little less than 10 seconds before the initial waves of the earthquake hit the city.

But for many earthquakes, the slower surface waves are the most destructive, since they tend to shake buildings and other structures with greater power. If the primary waves emanating from the earthquake trigger the alarm, citizens may have a half-minute or perhaps a little more to take action before the most violent events occur. Even 10 to 20 seconds is enough to find cover in most situations.

On October 1, 2007, Japan instituted an earthquake warning system administered by the Japan Meteorological Agency. An effective sys-

A technician at the Geotechnic Research Center in San Salvador, El Salvador, monitors seismometer recordings. *(Yuri Cortez/ AFP/Getty Images)*

tem needs enough sensors to cover a broad area so that there is a good chance an earthquake will be detected early; an earthquake arising between detectors situated too sparsely would travel a long time before being noticed, and the warning would come too late for people who lived nearby. Japan's system employs about 1,000 sensors buried underground. The sensors transmit information to a computer network that monitors and analyzes the data. A few seconds after an earthquake, the computer estimates the location and intensity and sends a warning to the affected region if the earthquake is serious enough to pose a threat.

The alarm goes out on the major television and radio channels. Instructions are also provided to prevent a panic, such as a massive rush for the exits in crowded buildings. Some of the response can be automated; for example, switches to shut off heavy machinery and elevators can be linked to the warning system. People may also have time to turn off gas lines, reducing the chance of a fire in their home or apartment. Fires are especially dangerous because the rubble and broken water lines make fighting fires exceptionally difficult in the aftermath of an earthquake. Much of the destruction of the 1906 San Francisco earthquake resulted from fires that burned out of control.

Seismologists and engineers in California are studying plans to build a similar warning system, but Japan's system is not yet perfect. Its first alarm, issued in April 2008 for an earthquake on the island of Okinawa having an approximate magnitude of 5.2, came a few seconds too late to provide adequate warning. The earthquake was minor, though, and did little damage.

Warning systems are also necessary for tsunamis. No alarms went out during the Indian Ocean tsunami of 2004, which caught everyone off guard. Undersea earthquakes cause these giant waves by disturbing huge masses of water, but not all undersea earthquakes generate tsunamis. Although it is not yet possible to tell from the seismic waves alone if an earthquake will cause a tsunami, detection of an undersea earthquake means that a tsunami is possible. Buoys and sensors to gauge the water level provide more specific signs of a tsunami and can alert people along low-lying coastal areas who are at risk. The Pacific and Atlantic Oceans have tsunami warning systems, and now the Indian Ocean does also.

Systems based on science and technology are extremely useful, but some observers claim that nature offers its own warning. For example, Sri Lanka (formerly Ceylon), an island nation off the southern tip of India, was in the path of the 2004 Indian Ocean tsunami, and Yala National Park was hit hard—yet few animal carcasses were found, suggesting that animals suffered fewer casualties than humans. Many other reports and observations through the years claim that certain animals behave strangely before the onset of seismic activity, as if they can sense the trouble ahead—or the ground shaking—much sooner than people can.

WATCHDOGS: ANIMAL BEHAVIOR PRECEDING AN EARTHQUAKE

What could the animals at Yala National Park have sensed to send them fleeing for high ground? It is possible that seismic events such as the undersea earthquake that caused the Indian Ocean tsunami may emit low-frequency waves called infrasound. Infrasound refers to sound waves having a frequency too low for most people to hear (*infra* Latin for "under" or "below"). Adult humans generally cannot hear sounds with a frequency lower than 20 hertz (although children and young adults often do a little better, hearing sounds down to about 15 hertz). But the hearing range of other animals extends into a lower range, and elephants communicate with sounds as low as 5–10 hertz, possibly by using their sensitive trunks or feet to detect low-frequency vibrations.

Vibrations of low frequency but high amplitude preceding the tsunami may have spooked the elephants at Yala National Park and in other affected areas, such as Thailand. Eyewitness reports on the day of the disaster describe agitated elephants fleeing from coastal areas. This behavior may have alerted other animals, or they may have also felt some kind of vibration.

Similar reports have linked agitated behavior of dogs, chickens, bees, rats, and many other creatures to impending earthquakes. Such reports have been made as long ago as 373 B.C.E., when rats and snakes supposedly fled their burrows prior to a severe earthquake in Greece. The USGS conducted some studies in the 1970s to test the possibility that animal behavior may predict earthquakes, but these studies did not find any evidence to support the claim. Some researchers in China and Japan continue to study animal behavior prior to earthquakes, but have not found any convincing proof.

But the reports continue. After the Sichuan earthquake on May 12, 2008, stories in the news described unusual behavior of animals in zoos or in the wild that had been noticed in the days leading up to the earthquake, including zoo elephants that swung their trunks and lions that paced their cages. The problem is that these reports almost always come after the fact. Some or even most of the reports may be true, but even so they are not scientifically significant. There are many reasons why animals may become agitated—for example, the presence or even just the scent of predators or prey would usually trigger a similar response. But people will often recall unusual animal behavior only when it precedes an earthquake or some other memorable event, not at other times, so such behavior becomes linked with earthquakes even though it happens frequently when there are no earthquakes. As for the animals at Yala National Park, they may have simply been able to outrun the waves, or if they failed to outrun them their carcasses may have washed out to sea, which would explain why few were found.

Despite the lack of scientific evidence for animal predictions of earthquakes, some animals do have extraordinary sensory powers. Certain snakes can sense infrared radiation, migrating birds and other animals can detect magnetic fields, and dolphins and bats use high-frequency sounds waves to find prey. Some animals may have special senses that, on certain occasions, might allow them to feel imminent earthquakes. The California Institute of Technology researcher Joseph L. Kirschvink suggests that such sensory systems may have evolved to give animals a greater chance of survival in earthquake-prone areas. In an article titled "Earthquake Prediction by Animals: Evolution and Sensory Perception," published in a 2000 issue of the *Bulletin of the Seismological Society of America,* Kirschvink proposed that animals may have developed a variety of sensory mechanisms to predict the onset of earthquakes. One of the most important mechanisms is related to vibrations or early movements that may herald the "main event."

SIGNS OF AN IMPENDING EARTHQUAKE

If certain types of animals are able to sense faint vibrations that may signal further seismic activity, then scientists with their sensitive seismometers should be able to do the same. The hypothesis is that these faint vibrations precede a major earthquake at least on some occasions, which means they can be used to predict when and where future earthquakes will occur. This idea is an active area of research in the field of earthquake prediction.

Earthquakes are well known to occur in clusters. The event of highest magnitude is sometimes called the main shock, while smaller earthquakes preceding it are called foreshocks and smaller earthquakes following it are called aftershocks. These minor earthquakes may occur hours, days, or weeks before or after the main shock. The difficulty with using foreshocks to predict large earthquakes is that small, random earthquakes occur quite frequently; hundreds of thousands of little earthquakes happen throughout each year, while only a few major earthquakes take place. A huge percentage of these low-magnitude earthquakes are obviously not associated with any high-magnitude earthquake.

Yet there may be reasons for believing that some type of seismic activity is common before major earthquakes hit. An analogous situation occurs with volcanoes, as magma flowing beneath the surface generally precedes major eruptions, and this flow generates weak seismic waves known as tremors. Many people use the term *tremor* as a synonym for earthquake, but geologists tend to use the term specifically to refer to seismic activity associated with the movement of magma. However, other kinds of activity may induce faint seismic pulses such as tremors. Suppose that as a fault starts to slip, it shudders or vibrates at a low frequency as more and more rocks start to crack under the strain. This vibration may emit low-amplitude seismic waves similar to tremors and would signal a major fault slippage and a potentially destructive earthquake.

Researchers have begun looking into this possibility. In 2002 Kazushige Obara, a researcher at the National Research Institute for Earth Science and Disaster Prevention (NIED) in Japan, recorded unusual tremors with NIED's network of about 600 seismic stations scattered throughout Japan. This sensitive network was designed to pick up small earthquakes, and Obara found tremors typically lasting a few tenths of a second up to a few seconds emanating from an area with no volcanic activity. The tremors seemed connected to the subduction—sliding underneath—of the Philippine Sea tectonic plate under the Eurasian plate that occurs in this part of Japan. Obara reported his finding in a *Sci-*

ence paper, "Nonvolcanic Deep Tremor Associated with Subduction in Southwest Japan." Researchers have also discovered nonvolcanic tremors in the Nankai Trough, a depression in the floor of the Pacific Ocean off the southwest coast of Japan created by subduction.

Where exactly are these tremors coming from? David R. Shelly (now working for USGS) and Gregory C. Beroza of Stanford University, along with Satoshi Ide and Sho Nakamula at the University of Tokyo in Japan, studied these seismic waves to determine their point of origin. Since the waves are weak and episodic, tracing them to their source was extremely difficult. After carefully analyzing more than 1,000 seismograms from the Nankai Trough recorded in 2002 to 2005, the researchers finally determined the arrival times of the waves and could trace the source. The source turned out to be a region of the fault between the two tectonic plates that has a pocket of fluid under high pressure. A lot of these waves had low frequencies. The researchers published their results, "Low-Frequency Earthquakes in Shikoku, Japan, and Their Relationship to Episodic Tremor and Slip," in a 2006 issue of *Nature*.

Shelly, Beroza, and Ide then analyzed the seismograms to try to determine what was causing the tremors. Shelly and his colleagues discovered that the time course of the tremors matched that of low-frequency earthquakes in the region. Tremors seem to be swarms of small-magnitude, low-frequency seismic events associated with movement along the fault, but extremely slow movement—almost silent, in terms of seismic activity, except for the faint vibrations. The researchers published their paper, "Non-Volcanic Tremor and Low-Frequency Earthquake Swarms," in a 2007 issue of *Nature*.

This faint, low-frequency activity could be the beginning of a major slippage, which means the tremors may prove important in earthquake prediction, or the events may be part of a normal creaking and groaning of a fault that has nothing to do with the next major earthquake. Perhaps this low-frequency activity intensifies immediately prior to a severe earthquake, and perhaps this activity is what agitates certain kinds of animals just before the earthquake—or perhaps not. An important goal of future seismological research is to address these issues.

Other researchers have based their efforts on the occurrence of seismic waves of higher amplitudes. These researchers study the past seismic activity of a region, as recorded on seismograms, to determine what activity may occur in the future. For example, foreshocks preceding a major earthquake may have particular patterns at particular faults, and if researchers can identify this pattern, they would be able to anticipate which minor earthquakes will probably be followed by a major one.

But past performance is not necessarily a predictor of the future (as stockbrokers and mutual fund managers often say). Most geologists use historical data to make forecasts rather than specific predictions. But earthquake forecasts, like stock market forecasts, often leave a lot to be desired.

EARTHQUAKE FORECASTS AND PROBABILITIES

Prediction is not yet possible, and people who live in earthquake-prone areas are apprehensive about the future. To keep these people as informed as possible—and without possibly misleading them with specific predictions in which experts presently do not have much confidence—Earth scientists make forecasts. These forecasts assess future hazards on the basis of what has happened, or failed to happen, in the past. For example, the 2008 report issued by USGS, California Geological Survey, and the Southern California Earthquake Center announced that California has more than a 99 percent chance of experiencing an earthquake of magnitude 6.7 M_w or greater in the next 30 years. The forecasters determined this probability by studying the past frequency of earthquakes in California as well as slip rates of the major faults in the state.

Several factors govern the probabilities of the 2008 report and other forecasts. Suppose, for instance, that an earthquake-prone area has been hit with six earthquakes of magnitude 6.7 M_w or better in the last 120 years. If those earthquakes occurred at random, then forecasters would anticipate an earthquake of magnitude 6.7 M_w or better to happen every 20 years or so.

But earthquakes of the past do not always strike at random intervals. The current theory of seismology, which holds that most earthquakes are due to the motion of tectonic plates as they grind or collide along faults, indicates that fault slippage is a critical component. When friction or rock protrusions arrest the motion along a fault, stress builds up, like a spring being wound tighter and tighter. Stress is finally released along the fault in a sudden spasm as the rocks crack or fail. This mechanism suggests that the severest earthquakes occur after a quiet period, accompanied by little or no seismic activity since the fault is not slipping. Then, after a period of years or decades of peace—and probably just as the residents begin to forget that earthquakes happen—a sudden rupture, perhaps preceded by tremors, generates a killer earthquake.

Normal slip rates for a given fault, averaged over time, can give geologists a clue about whether movement has suddenly stopped. Records of previous earthquakes, whether written down in recent history or suggested by geological studies of the ancient past, may also show an indicative pattern of gaps in activity, followed by a series of earthquakes, then another gap, followed by another series. Seismic activity in some parts of California has this kind of pattern.

What part of California will be affected in the earthquake forecasted in the 2008 report? Although forecasts can assign probabilities to different regions—for instance, the 2008 forecast indicated that Los Angeles has a 67 percent chance of an earthquake of magnitude 6.7 M_w or larger in the next 30 years and San Francisco has a 63 percent chance—no one knows for certain. The lack of certainty is reflected in the probabilities. Overall, California will almost certainly be struck by a major earthquake in the next three decades, at least according to the 2008 forecast, but forecasters can only make educated guesses as to where.

Earthquake forecasts keep people alert and cognizant of future hazards, but forecasts lack specifics. To do better than issuing probabilities that apply to time frames extending decades into the future, researchers must understand earthquakes and seismic activity much better. Specific earthquake predictions have already been made on occasion, with mixed success—and such predictions have consequences whether they are right or wrong.

PREDICTIONS AND CONSEQUENCES

A specific prediction states a future earthquake's time of occurrence, magnitude, and location. Given these three pieces of information, citizens in the danger zone can take appropriate action to protect their property as much as possible and then evacuate. Damage will be minimized and casualties will be few, if any.

The ability to make accurate earthquake predictions would confer enormous benefits, especially in those regions that are prone to earthquakes, such as California and Japan. Scientists as well as people with little scientific training are working toward this goal. Motivating some of these people, at least to a certain extent, is the prospect of making a great deal of money if a successful prediction scheme is patented and sold to governments or private institutions. As a result of these strong incentives, arising from both philanthropy and personal gain, a large number of prediction methods have been proposed. Some of these methods, such as those based on animal behavior, incorporate observations and make some attempt at scientific validity, while others defy logic and reason and will not be reviewed here. No method has proven adequate.

This is not to say that there have been no successful predictions. The best-known success came in 1975, when Chinese officials evacuated the city of Haicheng, with a population of roughly 1 million, in northeast China. In the months preceding the evacuation order, seismologists detected a distinct increase in minor earthquakes in the region, which scientists interpreted as foreshocks, and many animals exhibited strange behavior—rats and snakes appeared dazed and chickens would not stay in their coops. There were also reports of changes in surface elevation and water levels. On February 4, 1975, at 7:36 P.M. (local time), the day after the government issued the evacuation order, an earthquake of magnitude 7.3 on the Richter scale shook the area. Such a major earthquake would have caused tens of thousands of deaths, possibly even 100,000, but due to the timely evacuation, the city only suffered 2,041 fatalities and about 25,000 injured.

The success at Haicheng in 1975 may have given Chinese scientists confidence, but it did not last long. On July 28, 1976, at 3:42 A.M. (local time), an earthquake of magnitude 7.8 on the Richter scale hit the Chinese city of Tangshan. No warning signs had been noted and no alarms had been raised. Tangshan's population at the time of the earthquake was about the same as Haicheng's, but the residents of Tangshan were unprepared, and the fatalities numbered some 250,000.

The failure to warn Tangshan, just a year and a half after a brilliant success at Haicheng, highlights the difficulties of earthquake prediction and the spotty record of those who have thus far attempted it. In addition to unforeseen earthquakes, predictions have sometimes warned about earthquakes that failed to materialize, at least in the postulated time frame. In 1985 USGS warned the region around Parkfield, California, a small town in Monterey County, that an earthquake of magnitude 6.0 M_w was highly likely to occur between 1986 and 1993. The region had experienced a number of earthquakes of similar size in the recent past, the last (at the time of the prediction) coming in 1966. A 6.0 M_w earthquake did strike Parkfield, but the day was September 28, 2004, more than a decade after the warning had expired. The prediction did an excellent job on the magnitude—it was right on the money—but not so well on the time of the event.

Inaccurate predictions either give a false sense of security when they fail to predict an earthquake or raise a false alarm when they predict an earthquake that does not show up. The stakes are high.

Some researchers are pessimistic about ever developing a reliable method for predicting earthquakes. Similar to stock markets and weather systems, the dynamics underlying tectonic plate motion and associated earthquakes may be too complex and involve too many variables for precise predictions. In 1999 the journal *Nature* sponsored a debate over the question, "Is the reliable prediction of individual earthquakes a realistic scientific goal?" Several researchers weighed in. Robert J. Geller of Tokyo University summarized why earthquake prediction is so difficult: "The Earth's crust (where almost all earthquakes occur) is highly heterogeneous, as is the distribution of strength and stored elastic strain energy. The earthquake source process seems to be extremely sensitive to small variations in the initial conditions (as are fracture and failure processes in general). There is complex and highly nonlinear interaction between faults in the crust, making prediction yet more difficult." Geller offered a pessimistic appraisal: "In short, there is no good reason to think that earthquakes ought to be predictable in the first place."

Other researchers were less gloomy. Max Wyss of the University of Alaska acknowledged the difficulties, yet was hopeful over the long term because "there can be no doubt that a preparatory process to earthquake rupture exists (foreshocks demonstrate this), and I am confident that ingenious and resilient people, who will come after us and will be amused by this tempest in a teapot about the prediction of earthquakes, will eventually improve our ability to predict some earthquakes in favourable areas."

Perhaps scientists and engineers should invest more time and effort into construction techniques rather than prediction methods. After all, buildings and bridges must withstand earthquakes—they cannot be evacuated even if an accurate prediction formula is developed—and reinforced structures would be less likely to fall, which would greatly minimize casualties.

But some researchers continue to remain optimistic about the possibility of earthquake prediction. Faint rumblings or low-frequency vibrations, such as those studied by Shelly, Beroza, Ide, Nakamula, and other researchers, offer hope of improved prediction techniques. Other strategies include looking for signs of stress at faults that have been associated with previous earthquakes, which may indicate that pressure is starting to build—and that the clock is ticking until the next major event.

STRAIN ACCUMULATION AND SURFACE DEFORMATION

Tectonic plate movement exerts tremendous forces on rocks along the boundaries and at faults, where slabs of Earth's crust bump and grind past one another. Although rocks are extremely hard objects, the forces are so great that strain develops—a change in shape as the rocks bend or become compressed. These changes are measurable indicators of tectonic tension—and an earthquake waiting to happen.

As mentioned earlier, geologists measure fault slippage after an earthquake has occurred in order to gauge the magnitude of the event, and earthquake forecasters take into account fault slippage rates when estimating the chances of future events. Measuring strain is also possible, although the deformations tend to be slight. Rocks bulge, bend, or compress, but it does not take much beyond a small change in shape before the breaking point is reached. At this point, rupture and an earthquake ensue, followed by a quiet period as the strain accumulates once again.

In conjunction with its earthquake prediction at Parkfield in the 1980s, USGS performed extensive monitoring of this region. Parkfield is located along the San Andreas Fault. Since forecasters expected an earthquake, researchers at USGS and the California Geological Survey wanted to keep a close watch. Seismic instruments to monitor the area had been sparse prior to the prediction, but beginning in 1985 scientists established dozens of monitoring stations around a 15.6-mile (25-km) stretch of the San Andreas Fault near Parkfield. Instruments included seismometers to detect seismic waves, strain meters to measure rock deformation, creep meters to measure fault offset, and global positioning system (GPS) receivers to indicate precise positions. Although the earthquake came later than expected, when it finally arrived it was caught on tape, that is, recorded by the numerous instruments.

The data obtained with these instruments give scientists one of their best looks at earthquake processes and should help them better understand fault slips, particularly those that occur along the San Andreas Fault. But even more data is needed. To get a view closer to the focus of earthquakes around the Parkfield area, researchers have to dig into the fissure. As described in the following sidebar, the San Andreas Fault Observatory at Depth (SAFOD) is a project to study the processes deep within the fault that generate earthquakes.

Learning about fault slips and movements requires painstaking efforts, and not much data is available for most faults. Although the Parkfield and

San Andreas Fault Observatory at Depth (SAFOD)

The Parkfield earthquake prediction missed the mark by about 11 years, but the widely anticipated event drew the attention and instrumentation of about 100 earthquake researchers to the area. After nearly 20 years of experience at the San Andreas Fault near Parkfield, researchers decided to probe deeper. In the summer of 2002, researchers drilled a 1.4-mile (2.2-km) vertical hole near (but not at) the fault. Funding this work was the International Continental Scientific Drilling Program, a multinational effort to promote geological studies involving drilling. This hole was a pilot project, designed to test researchers' equipment and their capacity to conduct experiments at the fault itself. After its success, researchers began drilling a deeper hole in June 2004, to install instruments into and across the fault.

This project is called the San Andreas Fault Observatory at Depth. While much seismological research relies on instruments placed at the surface or not far below, this project aimed to monitor fault movement deep below the surface. The National Science Foundation (NSF), one of the

SAFOD projects should rectify this problem for at least part of the San Andreas Fault, other active regions are much more thinly instrumented. But recently geologists have circumvented some of the difficulties by partnering with NASA and getting a bird's-eye view (or rather a satellite's).

An important indication of stress is deformation. One of the advantages satellites provide is the ability to examine a broad area for any gradual shift in surface features. One of the tools satellites use to do this is called interferometric synthetic aperture radar. This imaging system bounces electromagnetic radiation from the surface and uses a property of the reflected waves known as interference to make precise measurements of surface elevation. Some systems have a sensitivity of about main government agencies that funds scientific research in the United States, provided funding for this phase of the project, along with USGS. After reaching a vertical depth of about 5,000 feet (1,520 m), the drillers angled the hole to about 55 degrees, heading toward the fault zone. On September 28, 2004, the long-anticipated earthquake struck, and the drill site got quite a shaking but sustained little damage. The project continued, and the hole crossed the fault zone at a vertical depth of about 10,000 feet (3,050 m).

During several phases of the project, researchers collected sample cores (cylindrical sections of rock cut out by a drill pipe). As geologists analyze the chemical composition and mechanical properties, they will learn more about the stresses to which rocks at the fault are subjected. This information is vital because ruptures in this area are the source of major earthquakes. In September 2007, the drill team finished up at the site by installing instrumentation at about 10,500 feet (3,200 m). The instruments included seismometers, accelerometers to measure sudden movement, tiltmeters to measure angle, and a pressure transducer. Over the coming years, data collected from this site will give geologists a glimpse of the mechanisms acting directly inside the fault.

0.04 inches (0.1 cm) under certain conditions, which means that if the surface moves by that amount, a satellite using this equipment can detect it by comparing its present location with previous images or maps.

Satellite imagery, along with data and tools discussed earlier in this chapter, have been integrated into an earthquake model program called QuakeSim. Researchers at NASA, the University of Colorado, and other institutions developed these computer simulations and algorithms to study fault mechanisms and associated earthquakes. Although the simulator does not predict earthquakes, it does identify regions where major earthquakes are expected. The researchers, led by John B. Rundle of the University of Colorado, described general approaches to modeling complicated systems occurring in geology and biology in a paper, "Self-Organization in Leaky Threshold Systems: The Influence of Near-Mean Field Dynamics and Its Implications for Earthquakes, Neurobiology, and Forecasting," in a 2002 issue of *Proceedings of the National Academy of Sciences.* The paper included a map of central and southern California that specified small areas likely to be struck by an earthquake of at least magnitude 5. The majority of earthquakes since the researchers issued this report have taken place in the identified locations.

Incorporating crustal deformation and other data from faults will improve earthquake forecasts as well as models and simulations. Shimon Wdowinski at the University of Miami in Florida and colleagues at the Scripps Institution of Oceanography compared recent models of faults such as the San Andreas Fault with precise measurements of the faults over the last two decades. Although the models were often accurate, there was a significant difference between fault movements in the models and the observed behavior at a specific section of the San Andreas Fault and a few other places. To eliminate this discrepancy, Wdowinski and his colleagues believe that the models need to include more data on crust strength and stress. The researchers published their report, "Diffuse Interseismic Deformation across the Pacific-North America Plate Boundary," in a 2007 issue of *Geology.*

Monitoring faults from the surface, inside deep holes, and from space will give researchers a great deal of valuable information on the structure, physics, and motion of these earthquake generators. The question is whether this data and its subsequent analysis will be sufficient to develop an accurate prediction method. The research described above indicates the increase in knowledge should result in improved earthquake forecasts, but specific predictions—time, date, place, and magnitude—may or may not be possible.

CONCLUSION

Centuries of research, beginning with observations of the ancient Greeks and probably long before, have led to a detailed though still incomplete understanding of earthquakes. Earthquakes strike much more often in some areas of the world than others, and scientists have traced most earthquakes to their origin—the faults along tectonic plate boundaries, where huge slabs of slowly moving rocks collide or grind past one another. By monitoring faults that are particularly earthquake prone, researchers have gained enough clues to permit reasonably accurate fore-casts, though these forecasts are probabilistic, lacking the certainty of a specific prediction.

New ideas are necessary before accurate earthquake predictions can be achieved, if they ever can be. One proposal involves monitoring the entire globe from space. In 2003, a panel of scientists led by Carol A. Ray-mond, Soren Madsen, and Wendy Edelstein of NASA's Jet Propulsion Laboratory issued a report, "Global Earthquake Satellite System," outlin-ing a plan to implement a satellite monitoring system. Some of the tools needed for this system to work, such as interferometric synthetic aperture radar, already exist, but the plan calls for further investment to develop lightweight steerable antennas, low-power radar electronics, and faster data processing systems for advanced surveillance and image analysis.

For the plan to be effective, NASA would need to launch a number of satellites dedicated to the mission. The required number depends on al-titude—higher altitudes enable a satellite to observe and monitor a larger region, but the costs of launch and maintenance increase with altitude (more energy is needed to place satellites in orbit and to communicate with them, due to the greater distances involved). A few satellites at high-altitude orbits could cover the globe, but a larger number of low-orbit satellites would be required for full coverage. At a minimum, two satel-lites at an altitude above 620 miles (1,000 km) are needed to make sure researchers could observe any surface on Earth in a 24-hour time frame.

Global satellite monitoring with sensitive radar would give geologists the capacity to study the strain and deformation preceding earthquakes anywhere in the world. This information could provide the means to un-derstand more fully the process by which stress builds up and is then relieved with an earthquake-producing rupture. Although different faults may follow slightly different rules, depending on the nature of the move-ment and the composition of the rocks, earthquake prediction for many or even most of the world's earthquake zones may become possible.

Worldwide coverage is necessary because earthquakes can strike out-side of well-known earthquake zones. For instance, at 4:37 A.M. on April 18, 2008, an earthquake of magnitude 5.2 M_w struck southern Illinois, close to the Indiana border. Although the region is far from any plate boundary, it is close to the Wabash Valley Fault, the earthquake's origin, which is an example of a fault within the body of a plate—an intraplate fault, most of which are not well understood. Residents escaped with only slight damage—this time.

At present, the global earthquake satellite system is not funded, and no start date has been set, since funding agencies are unsure if earthquake prediction is even feasible. The complexity of tectonic motion and faults, which has only been sketched in this chapter, may defeat any and all attempts at precise earthquake prediction. But no one can say for certain at this point. Perhaps people would be willing to settle for a system that is accurate most of the time, even though this means some earthquakes would occur unexpectedly and some false alarms would be raised. Considering the importance of advanced warning in minimizing damage and saving lives, at least a few scientists at the frontiers of Earth science will continue working toward this goal.

CHRONOLOGY

373 B.C.E.	People living in Greece notice peculiar behavior in animals prior to an earthquake.
1760 C.E.	The British scientist John Michell (1724–93) proposes that subsurface motions of rocks cause earthquakes.
1870s	Italian researchers such as Filippo Cecchi (1822–87) begin making early versions of seismometers.
1895	The Scottish-American geologist Andrew Lawson (1861–1952) discovers the northern portion of the San Andreas Fault.
1897	The German scientist Emil Wiechert (1861–1928) fashions a seismometer that is able to record throughout an earthquake episode (without breaking!).
1902	The Italian researcher Giuseppe Mercalli (1850–1914) designs a scale to measure earthquake intensity based on eyewitnesses and observational evidence.
1906	A powerful earthquake strikes San Francisco, California, damaging buildings and causing fires that destroyed much of the city and claimed about 3,000 lives.

1910	The American geologist Henry F. Reid (1859–1944) proposes the elastic rebound theory of earthquakes.
1931	The American seismologists Harry Wood and Frank Neumann modify the Mercalli scale, categorizing the intensity of earthquakes by Roman numerals from I to XII, with XII being the maximum.
1935	The American seismologist Charles Richter (1900–85) and the German-American seismologist Beno Gutenberg (1889–1960) develop an earthquake magnitude scale based on the amplitude of the vibrations as measured by their seismograph.
1975	Chinese scientists observe surface deformation and other signs of an impending earthquake and evacuate the city of Haicheng. A day later, an earthquake of magnitude 7.3 on the Richter scale strikes, but the city suffers few casualties due to the early warning.
1976	An earthquake of magnitude 7.8 on the Richter scale hits the Chinese city of Tangshan without warning, killing about 250,000 people in one of the worst natural disasters of the 20th century.
1979	The American seismologist Thomas Hanks and the Japanese seismologist Hiroo Kanamori introduce the moment magnitude scale for earthquake intensity, which most seismologists now prefer to use.
2007	Japan completes installation of a network of seismic sensors in the attempt to provide its citizens with at least a few seconds warning before a major earthquake hits.
	USGS staff finish placing seismic monitoring instruments deep within the San Andreas Fault, part of the San Andreas Fault Observatory at Depth project.
2008	USGS announces an earthquake forecast that warns of a 99 percent probability of an earthquake

of magnitude 6.7 M_w or greater in California in the next 30 years.

FURTHER RESOURCES

Print and Internet

Bolt, Bruce. *Earthquakes,* 5th ed. New York: W. H. Freeman, 2003. Excellent for students but not too technical, this book guides readers through the science of seismology.

Exploratorium Learning Studio. "Earthquakes." Available online. URL: http://www.exploratorium.edu/lc/pathfinders/earthquakes/. Accessed May 4, 2009. The Exploratorium, a museum of science, art, and perception in San Francisco, has maintained an excellent online presence since 1993. These pages discuss earthquake mythology, history, Richter scale, plate tectonics, seismology, and many other relevant topics.

Geology.com. "Eastern Sichuan, China Earthquake." Available online. URL: http://geology.com/events/sichuan-china-earthquake/. Accessed May 4, 2009. This article summarizes the Sichuan earthquake of May 12, 2008.

Global Earthquake Satellite System. "A 20-Year Plan to Enable Earthquake Prediction." March 2003. Available online. URL: http://solidearth.jpl.nasa.gov/GESS/3123_GESS_Rep_2003.pdf. Accessed May 4, 2009. This 2.8 megabyte file describes an ambitious plan to use satellite technology to develop an earthquake prediction method.

Guardian.co.uk. "Thousands Die in China Quake." May 13, 2008. Available online. URL: http://www.guardian.co.uk/world/2008/may/13/china.naturaldisasters. Accessed May 4, 2009. This article reports on the massive earthquake that struck China on May 12, 2008.

Hough, Susan Elizabeth. *Earthquaking Science: What We Know (and Don't Know) about Earthquakes.* Princeton, N.J.: Princeton University Press, 2002. This accessible book, written by a geophysicist, explains the current theory of earthquakes and offers a chapter on the ultimate goal of seismologists—earthquake prediction.

———. *Finding Fault in California: An Earthquake Tourist's Guide.* Missoula, Mont.: Mountain Press Publishing, 2004. If a person wants to see what a fault really looks like, then visiting California is a wise choice. This book will tell the reader how to find the faults and what to look for.

Jet Propulsion Laboratory. "QuakeSim." Available online. URL: http://quakesim.jpl.nasa.gov/. Accessed May 4, 2009. QuakeSim is project of JPL—a NASA research center—incorporating data from satellite imagery, seismometers, and previous earthquake frequencies into a computerized earthquake model to identify areas of potential seismic activity. This Web resource gives the details and an updated scorecard to determine how well they are doing.

Kirschvink, Joseph L. "Earthquake Prediction by Animals: Evolution and Sensory Perception." *Bulletin of the Seismological Society of America* 90 (2000): 312–323. The author describes what is known about the study of animal behavior to predict earthquakes and hypothesizes about the mechanisms animals could possibly use to sense early signs of earthquakes.

National Aeronautics and Space Administration. "Global Earthquake Satellite System." Available online. URL: http://solidearth.jpl.nasa.gov/gess.html. Accessed May 4, 2009. The goal is to develop a satellite-based system to monitor the entire world for surface strain or deformation that could predict the sites of future earthquakes. A link to a 104-page report is included.

Nature.com. "Is the Reliable Prediction of Individual Earthquakes a Realistic Scientific Goal?" February 25, 1999. Available online. URL: http://www.nature.com/nature/debates/earthquake/equake_frameset.html. Accessed May 4, 2009. *Nature* sponsored an open debate to discuss the possibility of developing accurate methods of earthquake prediction.

Obara, Kazushige. "Nonvolcanic Deep Tremor Associated with Subduction in Southwest Japan." *Science* 296 (May 31, 2002): 1,679–1,681. Obara recorded unusual tremors in a network of about 600 seismic stations scattered throughout Japan and found tremors typically lasting a few tenths of a second up to a few seconds that emanated from an area with no volcanic activity.

Rundle, J. B., K. F. Tiampo, W. Klein, and J. S. Sá Martins. "Self-Organization in Leaky Threshold Systems: The Influence of Near-Mean Field Dynamics and Its Implications for Earthquakes, Neurobiology, and Forecasting." *Proceedings of the National Academy of Sciences* 99

(February 19, 2002): 2,514–2,521. The researchers describe general approaches to modeling complicated systems occurring in geology and biology.

ScienceDaily. "California Has More Than 99% Chance of a Big Earthquake Within 30 Years, Report Shows." News release, April 15, 2008. Available online. URL: http://www.sciencedaily.com/releases/2008/04/080414203459.htm. Accessed May 4, 2009. This news release summarizes a report of the USGS, California Geological Survey, and Southern California Earthquake Center.

Shelly, David R., Gregory C. Beroza, Satoshi Ide. "Non-Volcanic Tremor and Low-Frequency Earthquake Swarms." Nature 446 (March 15, 2007): 305–307. The researchers discovered that the time course of certain tremors match that of low-frequency earthquakes in the region.

Shelly, David R., Gregory C. Beroza, Satoshi Ide, and Sho Nakamula. "Low-Frequency Earthquakes in Shikoku, Japan, and Their Relationship to Episodic Tremor and Slip." Nature 442 (July 13, 2006): 188–191. The researchers studied more than 1,000 seismograms from the Nankai Trough recorded in 2002–05 and traced the source of low-frequency earthquakes to a region of a fault between two tectonic plates that has a pocket of fluid under high pressure.

Ulin, David L. The Myth of Solid Ground: Earthquakes, Prediction, and the Fault Line Between Reason and Faith. New York: Viking, 2004. The ability to predict earthquakes has eluded even the brightest geologists, but the importance of making accurate predictions has encouraged many people, including those with little training or scientific acumen, to give it a shot. Ulin introduces the subject, along with the scientific—and sometimes unscientific—methods and theories.

United States Geological Survey. "Earthquake Hazards Program." Available online. URL: http://earthquake.usgs.gov/. Accessed May 4, 2009. An important task of USGS is to monitor earthquakes across the globe. This Web resource describes this process as well as providing a ton of basic information on earthquake and earthquake research, including a section on earthquake prediction.

———. "San Andreas Fault." Available online. URL: http://pubs.usgs.gov/gip/earthq3/contents.html. Accessed May 4, 2009. These pages describe this long, prominent fault of California and its important seismological properties.

———. "San Andreas Fault Observatory at Depth." Available online. URL: http://earthquake.usgs.gov/research/parkfield/safod_pbo.php. Accessed May 4, 2009. The USGS offers a description of the San Andreas Fault Observatory at Depth (SAFOD) project, including an overview and site characterization studies.

———. "USGS Earthquake Magnitude Policy." Available online. URL: http://earthquake.usgs.gov/aboutus/docs/020204mag_policy.php. Accessed May 4, 2009. USGS describes its policies regarding the techniques used to measure magnitude.

Virtual Museum of the City of San Francisco. "1906 Earthquake and Fire." Available online. URL: http://www.sfmuseum.org/1906/06.html. Accessed May 4, 2009. Time lines, historic newspaper clippings, eyewitness accounts, lists of casualties, fire and police department reports, engineering and scientific reports, relief efforts, and a large collection of photographs tell the story of the disaster and its aftermath.

Wdowinski, Shimon, Bridget Smith-Konter, Yehuda Bock, and David Sandwell. "Diffuse Interseismic Deformation across the Pacific-North America Plate Boundary." *Geology* 35 (2007): 311–314. The researchers compared recent models of faults such as the San Andreas Fault with precise measurements of the faults over the last two decades.

Winchester, Simon. *A Crack in the Edge of the World: America and the Great California Earthquake of 1906*. New York: HarperCollins, 2005. The disastrous 1906 earthquake left San Francisco mostly rubble, as buildings collapsed and fires raged out of control. Winchester revisits the carnage and discusses the subsequent geological investigation of the San Andreas Fault.

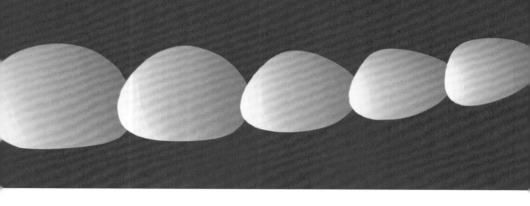

FINAL THOUGHTS

Of all the puzzles at the frontiers of Earth science, perhaps the most intriguing is how life arose on the planet. Plants and animals reproduce their own kind, though sometimes with variations, as the British naturalist Charles Darwin (1809–82) recognized when he proposed the theory of evolution. This process explains how life maintains itself and evolves over time, but it does not explain how it got started in the first place.

Where and how life originated is a fascinating topic for many reasons and has special meaning because the solution to the mystery will reveal much about the nature of life, including the species known as *Homo sapiens*—modern humans. It is also a controversial topic, since the genesis and nature of life, particularly *Homo sapiens,* is also the province of religion. The author of this book has no intention of denigrating any form of religious belief, but scientific evidence, including studies of radioactive isotopes, clearly indicates that Earth is about 4.5 billion years old, and fossils document the evolution of life on the planet.

No one knows exactly when life began on Earth. Ancient fossils are difficult to find, and the fossil record that scientists have uncovered to date is incomplete. And since early organisms were tiny, researchers have trouble distinguishing the remains of ancient life from the artifacts of chemical or geological processes. The oldest fossils are presently the subject of debate, but relics of ancient microbes have been discovered in rocks as old as 3 billion years. Life began early in Earth's history—at least 3 billion years ago, and possibly quite earlier.

Billions of years ago, Earth was much different than it is today. Geologists believe the atmosphere during this time came from volcano outgassing, which would have contained carbon dioxide, nitrogen, hydrogen, water vapor, and a few other substances. Hydrogen is so light that some of it escaped, and some of it combined with carbon or nitrogen to form methane or ammonia, among other hydrogen-containing compounds. Oxygen came later, as a by-product of the photosynthesis of plants.

The exact composition of the early atmosphere is unknown, but in 1953 Stanley Miller (1930–2007), a graduate student studying under Professor Harold Urey (1893–1981) at the University of Chicago, reported an astonishing experiment in which he assumed the early atmosphere had abundant hydrogen compounds. (Hydrogen-rich atmospheres are called reducing atmospheres, since their chemistry would involve reducing reactions, as opposed to oxidizing reactions.) Miller put water, hydrogen, methane, ammonia, and carbon monoxide in a sterilized flask and then exposed the contents to electric sparks that simulated lightning. The water turned brown a few weeks later. Analyzing the contents, he found a large number of organic—carbon-containing—compounds, including several different types of amino acids, the building blocks of proteins.

The experiments of Miller and Urey suggested that molecules critical for life could have developed from simple reactions in a reducing environment. Although the experiments did not create life itself, they suggested how the vital components could have arisen. Interactions among these components presumably led to life, some time in the distant past.

But Earth may not have had such a reducing atmosphere in its early days. Some researchers have returned to the experiments of Miller and Urey, but with the focus on a specific instigator—volcanic activity.

Adam P. Johnson, a graduate student with the NASA Astrobiology Institute at Indiana University, and Jeffrey L. Bada at the Scripps Institution of Oceanography came across some material left over from Miller's 1950s experiments. Miller had moved to the University of California, San Diego, in 1960, and Bada was a graduate student in his laboratory in 1965–68. In an interview for a NASA press release on October 16, 2008, Bada said, "Stanley and I continued to work on various projects until he died in 2007. When Adam and I found the samples from the original experiments, it was a great opportunity to reanalyze these historic samples using modern methods."

In some of the experiments, Miller had simulated volcanic eruptions by injecting steam into the apparatus. Johnson, Bada, and their colleagues focused on these experiments because they felt that conditions during a volcanic eruption would be most likely to resemble the gaseous composition that Miller had chosen. Miller had found five different kinds of amino acids when he analyzed the results of these experiments, but Johnson, Bada, and their colleagues found 22 amino acids when they used the more sensitive instruments available today. The amount of amino acids these volcanic experiments produced was at least as great and sometimes greater than in other experimental arrangements. Johnson, Bada, and their colleagues published their findings, "The Miller Volcanic Spark Discharge Experiment," in a 2008 issue of *Science.*

The focus on volcanoes is important, because volcanoes emit a considerable amount of hydrogen. In their paper, the researchers noted, "Geoscientists today doubt that the primitive atmosphere had the highly reducing composition Miller used. However, the volcanic apparatus experiment suggests that, even if the overall atmosphere was not reducing, localized prebiotic synthesis could have been effective." In other words, volcanoes and their emissions would have provided the essential conditions in which molecules critical to the development of life could have formed.

The results of this study mark an important step toward the eventual solution to the question of life's origins, but nobody can be certain in what direction future experiments and theory will go. Life began under special circumstances and at a slow speed, at least in terms of human lifetimes, so replicating the entire process in a laboratory is not possible at the present time. And a new form of life would be highly unlikely to arise these days, due to Earth's altered conditions and the presence of so much preexisting life that would eat it.

Some people propose that life did not begin on Earth at all, but came from space in some form, perhaps hitching a ride on meteorites. But this theory, called panspermia, merely extends the mystery to another place, for one can ask how life arose there as well. Since there is no convincing evidence that life came from elsewhere, the simplest explanation at present is that life on Earth originated here on the planet.

The rise of life, in which large, organic molecules seemed to have pulled themselves up "by their bootstraps" into living organisms, is a perplexing question at the frontier of science. Progress toward a solution of this mystery would help people learn a great deal about the nature of living organisms. Perhaps life is an inevitable process, which takes place almost everywhere it can get a foothold, in which case the search for life beyond Earth will one day be successful. Or perhaps life on Earth is due to a juxtaposition of so many unlikely events that Earth is the only inhabited planet in the universe.

Research into life's origins may reveal not only how life got started, but under what conditions it could possibly end. Considering the global changes Earth is currently experiencing, this issue might be the most relevant of all.

GLOSSARY

aquifer   underground area that has enough water to supply a well

asthenosphere   a weak layer of rock in the upper mantle

caldera   a basin or sink formed by the collapse of a volcano

chron   *See* **magnetic chron**

compression waves   propagating disturbances in which material alternately squeezes (compresses) and pulls apart

convection   mechanism of heat transfer by which flows of material such as liquid or molten rock carry heat from warm to cool regions

crust   thin layer covering Earth's surface, extending from continents to an average depth of about 22 miles (35 km) and from the ocean floor to an average depth of about four miles (6.4 km)

current   in electricity, a flow of electric charges

desalination   removal of salt and minerals from seawater in order to render it drinkable

dormant   describes a volcano that has been inactive for a considerable period of time

electrons   negatively charged particles normally found in motion around an atom's nucleus

epicenter   in seismology, the ground directly above an earthquake's focus

fault break or crack in rocks where one side has moved relative to another

focus in seismology, the site of an earthquake's origin, or initial disturbance, which is usually located under the surface and propagates from this point in all directions

fossil fuels energy resources—oil, coal, and natural gas—formed long ago from the remains of organisms

hydrology the study of the properties, distribution, and circulation of water

inner core Earth's solid center, composed of iron with a little nickel, with a radius of about 756 miles (1,220 km)

ions charged particles

isotopes atoms of the same element but with different numbers of neutrons

lava molten rock that erupts from a volcano

lithosphere crust and uppermost mantle to a depth of about 60 miles (100 km)

M_w moment magnitude scale (sometimes abbreviated as M)

magma molten rock beneath Earth's surface

magnet an object having the capacity to exert an attractive force on iron in its vicinity

magnetic chron period in which Earth's magnetic field is relatively stable at one orientation or the other

magnetic field a region of space in which a magnet exerts its forces

magnetic pole one of the ends of a magnet

magnetometers instruments designed to measure magnetic fields

magnetosphere the magnetic field surrounding a planet or star and extending into space

mantle the rocky region of Earth's interior extending from the crust to about 1,800 miles (2,900 km) below the surface

Moho *See* **Mohorovicic discontinuity**

Mohorovicic discontinuity boundary between the crust and mantle

moment magnitude a measure of the energy of an earthquake, based on the movement of the rocks at the origin

outer core hot, liquid region of Earth's interior, composed mostly of iron and nickel, extending from the mantle to a depth of about 3,200 miles (5,150 km)

P waves *See* **primary waves**

paleomagnetism record of Earth's magnetic field as stored in the structure of old rocks

plumes channels of exceptionally hot material within Earth's crust and mantle

potable suitable for drinking

precipitation in weather systems, a fall of water of some form—rain, snow, or ice

primary waves fast seismic waves that are first to arrive at a given point from an earthquake; these waves are a type of compression wave

radioactivity emission of energetic particles from the nucleus of unstable atoms

relative humidity amount of water vapor in the air, given as a percentage of the maximum amount of water vapor that air can hold at the same temperature

Richter scale a system of judging the magnitude of an earthquake from the size of its seismic waves; developed in 1935 by Charles F. Richter (1900–85) and Beno Gutenberg (1889–1960)

S waves   *See* **secondary waves**

saturated zone   subsurface area in which all the pores in the rock and soil are filled with water

secondary waves   the second waves to arrive at a given point from an earthquake; these waves are a type of shear wave and do not propagate through fluid

sedimentary rock   formed when material such as sand, mud, and calcium carbonate shells is buried and compressed

seismic waves   vibrations in the ground produced by earthquakes

shear waves   disturbances in which the displacement is perpendicular to the direction of propagation

SQUID   *See* **superconducting quantum interference device**

strain   deformation caused by applied forces

superconducting quantum interference device   an instrument that is highly sensitive to magnetic fields and can detect and measure exceptionally weak fields

tectonic plates   large slabs of lithosphere that glide slowly along Earth's surface

United States Geological Survey   government agency established in 1879 to conduct mapping and geological studies

vadose zone   the portion of the underground that is dry or partially dry and exists between the surface and the saturated zone

water table   the boundary between the saturated zone and the vadose zone; in other words, it is approximately the "surface" of the underground water

FURTHER RESOURCES

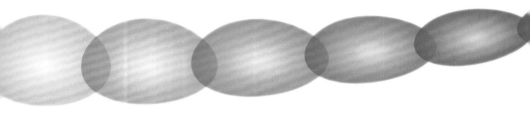

Print and Internet

Allaby, Michael, Robert Coenraads, et al. *The Encyclopedia of Earth: A Complete Visual Guide.* Berkeley: University of California Press, 2008. This reference work offers spectacular pictures of Earth's varied geology and landscapes, along with simple explanations of many natural phenomena. Main sections of the book discuss Earth's history, the dynamics of the interior, rocks and minerals, weather and climate, water, and the ways the planet and its geology have affected human society.

Forget, François, François Costard, and Philippe Lognonné. *Planet Mars: Story of Another World.* Berlin: Praxis Publishing, 2008. This book is a translation of a 2006 volume that discusses what scientists have learned about the evolution and geology of Mars. Although colder and more distant from the Sun than Earth, Mars has a fascinating structure, in some cases exhibiting features similar to terrestrial geology.

Fortey, Richard. *Earth: An Intimate History.* New York: Vintage, 2005. Fortey, a scientist at the Natural History Museum in London, surveys the planet and shows how geological processes have shaped the landscape. Readers get a guided tour of Vesuvius, the Hawaiian Islands, the Grand Canyon in Arizona, and many other fascinating sites that display and exemplify a great deal of Earth's evolution.

Johnston, Andrew K. *Earth from Space.* Buffalo, N.Y.: Firefly Books, 2004. Satellites have played a crucial role in geology, letting researchers study broad swaths of the planet with the flick of a switch. Johnston, a geographer in the research department of the National Air and Space Museum, selected 300 satellite images that provide a beautiful, global view of the planet.

Lewis, Cherry. *The Dating Game: One Man's Search for the Age of the Earth*. Cambridge: Cambridge University Press, 2002. Arthur Holmes is not well known in the annals of the history of science, but his contributions to the determination of Earth's true age make a fascinating story. In the early 20th century, when many scientists believed Earth was only a few million years old, Holmes's study of radioactivity derived a much more accurate age of several billion years.

Lomborg, Bjørn. *The Skeptical Environmentalist: Measuring the Real State of the World*. Cambridge: Cambridge University Press, 2001. Lomborg, a professor at the University of Aarhus in Denmark when he wrote this book, gained much notoriety when he criticized the global warming researchers who believe Earth's climate will continue to change rapidly and perilously. Lomborg's opinion is that the dire warnings of these researchers are exaggerated. People who summarily dismiss skeptics such as Lomborg would do well to keep this in mind: Contrarian views, while often wrong, can contain important truths that are missed by researchers who jump on whatever bandwagon happens to be popular at the moment.

McPhee, John. *Annals of the Former World*. New York: Farrar, Straus and Giroux, 1998. McPhee is a journalist and gifted writer who accompanied a group of geologists on a series of tours across the United States. This book describes the many geological features he saw and how geologists explain them. Even ordinary-looking rocks, exposed when highway workers cut a path for the road, have an interesting story to tell about Earth's past and present.

National Aeronautics and Space Administration. "Welcome to the Planets." Available online. URL: http://pds.jpl.nasa.gov/planets/. Accessed May 4, 2009. NASA geologists have the whole solar system to study, including Earth. This Web resource includes images and profiles of the planets and other bodies (including Pluto) and information on space missions such as Mars Global Surveyor and *Voyager*.

Repcheck, Jack. *The Man Who Found Time: James Hutton and the Discovery of Earth's Antiquity*. Cambridge, Mass.: Perseus

Publishing, 2003. Hutton, an 18th-century Scottish farmer and naturalist, is one of the founders of modern geology. Most people in the 18th century were convinced that Earth was young, but Hutton's observations of rock formations and other geological features, and the theories with which he explained them, suggested that the planet was vastly older. Hutton's ideas, as those of Galileo before him and Darwin afterward, were crucial strides in the progress of science.

Silver, Jerry. *Global Warming and Climate Change Demystified.* New York: McGraw-Hill, 2008. The "Demystified" line of books aims to explain a complex topic as simply and accurately as possible. This book, written by a science teacher, discusses the data, research techniques, and hypotheses of global climate change.

Space.com. "All About the Planets." Available online. URL: http://www.space.com/planets/. Accessed May 4, 2009. Links to images, articles, news, and research on the planets are collected on this page. Comparisons of Earth with the other bodies of the solar system (and beyond) helps geologists to understand planetary structure and evolution.

University of California Museum of Paleontology. "Tour of Geologic Time." Available online. URL: http://www.ucmp.berkeley.edu/exhibits/geologictime.php. Accessed May 4, 2009. The tour starts from Earth's beginning, about 4.5 billion years ago, and provides information on the geology and life-forms that existed at any given time in the planet's history.

U.S. Environmental Protection Agency. "Climate Change." Available online. URL: http://www.epa.gov/climatechange/. Accessed May 4, 2009. The EPA's Web site on global warming and climate change provides information on basic issues, the science of climate change, greenhouse gas emissions, health and environmental effects, climate economics, and climate policies and regulatory initiatives of the U.S. government.

U.S. Geological Survey. "Geology Research and Information." Available online. URL: http://geology.usgs.gov/. Accessed May 4, 2009. USGS, the government agency involved in mapping and geological studies, maintains this collection of data and news of the latest research.

———. "The USGS and Science Education." Available online. URL: http://education.usgs.gov/. Accessed May 4, 2009. Links to a wide variety of educational resources can be found here. Categories include primary grade (K–6) resources, secondary grade (7–12) resources, publications, maps, images, videos, and more.

Weisman, Alan. *The World Without Us.* New York: Thomas Dunne Books, 2007. Human civilization has left its mark on the planet, but it is not an indelible one, and it is fascinating to consider what would happen if the world was left to its own devices once again. (In other words, what would happen if humans become extinct.) This book describes the subsequent decay of buildings, roads, and other structures and how this decay would affect the planet and the future course of life.

Web Sites

Exploratorium. Available online. URL: http://www.exploratorium.edu/. Accessed May 4, 2009. The Exploratorium, a museum of science, art and human perception in San Francisco, has a fantastic Web site full of virtual exhibits, articles, and animations, including much of interest to geologists and geologists-to-be.

Geology.com. Available online. URL: http://geology.com. Accessed May 4, 2009. This site covers the whole spectrum of geology, from practical engineering issues such as oil drilling, to theoretical concerns such as the cause of tsunamis. Sections include news, careers, articles, maps, satellite images, and a dictionary of geological terms.

How Stuff Works. Available online. URL: http://www.howstuffworks.com/. Accessed May 4, 2009. This Web site hosts a huge number of articles on all aspects of technology and science, including geology.

ScienceDaily. Available online. URL: http://www.sciencedaily.com/. Accessed May 4, 2009. An excellent source for the latest research news, ScienceDaily posts hundreds of articles on all aspects of science. The articles are usually taken from press releases issued by the researcher's institution or by the journal that published the research.

Main categories include Fossils & Ruins, Mind & Brain, Earth & Climate, Matter & Energy, and others.

U.S. Geological Survey. Available online. URL: http://www.usgs.gov/. Accessed May 4, 2009. The Web pages of USGS contain an enormous quantity of excellent information and resources, including maps, imagery, seismology, earth science, geography, and much else.

INDEX

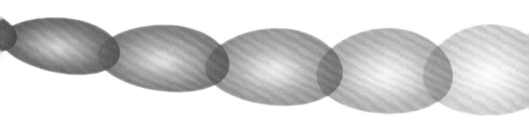